Nonlinear analysis and mechanics: Heriot-Watt Symposium
VOLUME II

R J Knops (Editor)
Heriot-Watt University

Nonlinear analysis and mechanics: Heriot-Watt Symposium

VOLUME II

Pitman
LONDON · SAN FRANCISCO · MELBOURNE

PITMAN PUBLISHING LIMITED
39 Parker Street, London WC2B 5PB

FEARON-PITMAN PUBLISHERS INC.
6 Davis Drive, Belmont, California 94002, USA

Associated Companies
Copp Clark Ltd, Toronto
Pitman Publishing New Zealand Ltd, Wellington
Pitman Publishing Pty Ltd, Melbourne

First published 1978

AMS Subject Classifications: (main) 73-XX 35-XX
(subsidiary) 73-46, 73-49, 35G25, 35G30

Reproduced and printed by photolithography
in Great Britain at Biddles of Guildford.

ISBN 0 273 08420 8

Preface

This second volume, as with others in the series, is devoted to written versions of lectures presented at Heriot-Watt University during part of a research programme into 'Qualitative Properties of Nonlinear Elasticity'. The programme is mainly sponsored by the Science Research Council, although it is a pleasure to acknowledge support also from the Carnegie Trust for the Universities of Scotland for the research fellowship held by J. E. Marsden.

The two articles comprising the present volume will be of value to those interested in current mathematical developments in nonlinear elasticity and, by implication, in the broader field of continuum physics.

The authors of the articles have willingly cooperated in the preparation of their manuscripts. The typing has been done admirably, often under severe conditions, by Mrs. M. E. Crawford. Her cheerful and patient persistence deserves special praise. Thanks are also due to Lynda Robertson for skilfully preparing the diagrams.

Edinburgh
June 1978

R. J. Knops

Contents

S S ANTMAN AND H BREZIS

The existence of orientation-preserving deformations in nonlinear elasticity

1 INTRODUCTION

Let B be the reference configuration of a body. We identify B with a region of $\mathbb{R}^3$ and we identify particles of the body with their coordinate triples $\underset{\sim}{X}$ in the reference configuration. Let $\underset{\sim}{p}(\underset{\sim}{X},t)$ denote the position of particle $\underset{\sim}{X}$ at time t. In continuum mechanics, $\underset{\sim}{p}$ is found as the solution of a system of equations. To prevent two distinct particles from simultaneously occupying the same point in space, $\underset{\sim}{p}(\cdot,t)$ must be a one-to-one mapping of B for each t.

A satisfactory treatment of the existence of globally invertible solutions of the governing equations of any field of nonlinear continuum mechanics seems beyond the scope of the available methods of analysis and topology. Consequently one is led to consider the far less ambitious problem of demonstrating that the governing equations possess orientation-preserving solutions, i.e., position fields $\underset{\sim}{p}$ satisfying

$$\det(\partial\underset{\sim}{p}/\partial\underset{\sim}{X}) > 0 \tag{1.1}$$

in a suitable sense. This problem is itself most formidable. We interpret (1.1) as a material constraint by requiring some stress to become unbounded as $\det(\partial\underset{\sim}{p}/\partial\underset{\sim}{X}) \longrightarrow 0$.

In this paper we shall examine this problem in the context of nonlinear elasticity. This theory presents the additional difficulty of finding restrictions on the constitutive functions that are physically reasonable, ensure existence of solutions, and yet allow nonuniqueness for certain statical problems. (Cf. Truesdell [20], Wang and Truesdell [22], Ch. III). Only recently has this problem been successfully attacked by Ball [10] and by Hughes, Kato and Marsden [14]. Ball showed that the potential energy functional for boundary value problems for hyperelastic bodies possesses a minimum in a class of orientation-preserving deformations. For this purpose he assumed that the strain energy function satisfies a condition somewhat more stringent that the strong ellipticity (strong Legendre-Hadamard) condition. Ball, however, could show that such minimizers are weak solutions of the Euler-Lagrange equations, which are the equilibrium equations for hyperelastic bodies, only when the restriction (1.1) is suspended. (Indeed, that minimizers of such problems are weak solutions of the Euler-Lagrange equations has only been demonstrated for the case in which these equations are ordinary differential equations. Cf. Antman [3], [4], [5], [8].) Hughes, Kato and Marsden [14] showed that the Cauchy problem for the quasilinear hyperbolic equations of nonlinear elastodynamics possess a solution for small time if the material satisfies the strong ellipticity condition. The local character of these results avoids difficulties with (1.1).

In this paper we examine a variety of strategies that are successful in proving the existence of orientation-preserving deformations of a model problem describing the deformation of a nonlinearly elastic rod by means of a system of ordinary differential equations. Some of these methods may prove useful in treating three-dimensional problems. We severely restrict the geometric structure, material behavior, nature of loads, boundary conditions, and regularity of data for our problem in order to bring the central idea into sharp focus. The problem we examine is, nevertheless, of considerable importance and retains far more generality than those commonly examined in mathematical treatments of structures. About half of our development represents simplifications of other work; the remainder, including Section 4, is new.

2 BOUNDARY VALUE PROBLEM FOR THE PLANAR DEFORMATION OF RODS

Let $(\underset{\sim}{i},\underset{\sim}{j},\underset{\sim}{k})$ be an orthonormal basis for the Euclidean 3-space $\mathbb{E}^3$. We consider the deformation of a two-dimensional body whose reference configuration is the rectangle $\{(s,y) \in \mathbb{R}^2 : s \in [0,1], y \in [-h,h]\}$. We assume that there are functions

$$s \in [0,1] \longmapsto (\underset{\sim}{r}(s),\underset{\sim}{b}(s)) \in \text{span}\,\{\underset{\sim}{i},\underset{\sim}{j}\}, \quad |\underset{\sim}{b}(s)| = 1 \tag{2.1}$$

such that the position $\underset{\sim}{p}(s,y)$ of the particle (s,y) in some arbitrary configuration is constrained to have the form

$$\underset{\sim}{p}(s,y) = \underset{\sim}{r}(s) + y\underset{\sim}{b}(s). \tag{2.2}$$

In this case we call the body a <u>rod</u>. <u>Its configuration</u> is defined by the functions $\underset{\sim}{r}$ and $\underset{\sim}{b}$.

We denote differentiation with respect to s by a prime. We define $\underset{\sim}{a}$ and θ by

$$\underset{\sim}{a}(s) \equiv \cos\theta(s)\underset{\sim}{i} + \sin\theta(s)\underset{\sim}{j}, \tag{2.3a}$$

$$\underset{\sim}{b}(s) \equiv -\sin\theta(s)\underset{\sim}{i} + \cos\theta(s)\underset{\sim}{j}. \tag{2.3b}$$

(See Fig. 1.1) We define strains ξ and η by

$$\underset{\sim}{r}'(s) \equiv [1 + \xi(s)]\underset{\sim}{a}(s) + \eta(s)\underset{\sim}{b}(s) \tag{2.4}$$

We set

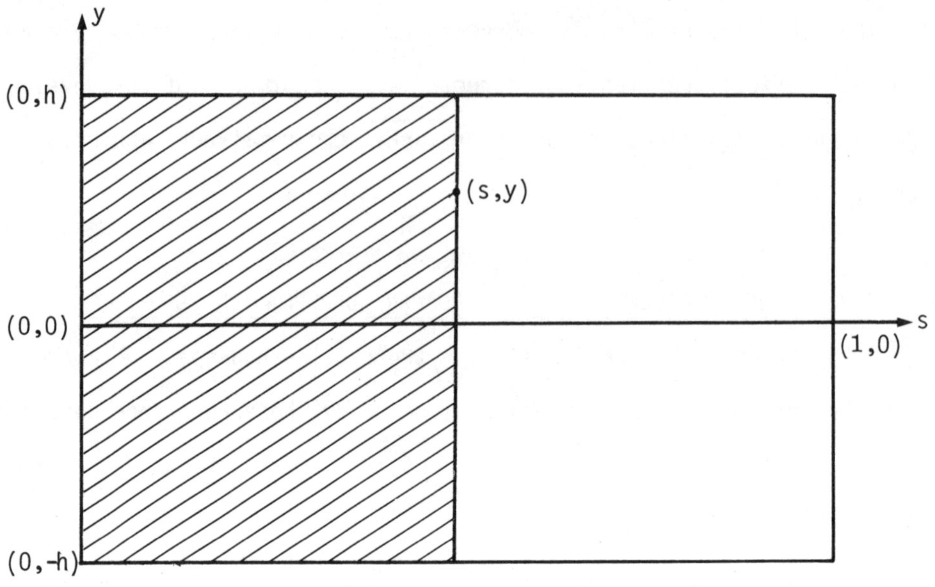

Reference Configuration

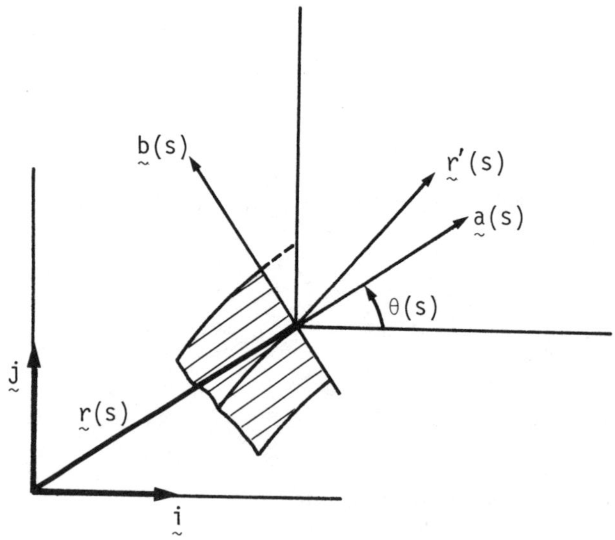

Deformed Configuration

Fig. 1.

$$\mu(s) \equiv \theta'(s). \tag{2.5}$$

The triple of functions

$$\underset{\sim}{u} \equiv (\xi,\eta,\mu) \tag{2.6}$$

are the <u>strains</u> for our problem.

The requirement (1.1) that $\underset{\sim}{p}$ preserve orientation reduces to

$$\underset{\sim}{k}\cdot\left(\frac{\partial \underset{\sim}{p}}{\partial s} \times \frac{\partial \underset{\sim}{p}}{\partial y}\right) > 0 \qquad \forall\ (s,y) \in [0,1] \times [-h,h]. \tag{2.7}$$

This is equivalent to the inequality

$$\delta(\underset{\sim}{u}) \equiv 1 + \xi - h|\mu| > 0. \tag{2.8}$$

(Conditions for the preservation of orientation in far more sophisticated models have been developed by Antman [7]). Thus (2.8) implies that orientation-preserving strains $\underset{\sim}{u}$ are those whose values lie in the <u>open</u> convex set

$$K \equiv \{\underset{\sim}{u} : \delta(\underset{\sim}{u}) > 0\}. \tag{2.9}$$

We choose the basis $(\underset{\sim}{i},\underset{\sim}{j})$ and the origin $\underset{\sim}{0}$ such that

$$\underset{\sim}{r}(0) = \underset{\sim}{0}, \qquad \theta(0) = 0. \tag{2.10}$$

Then (2.4) and (2.5) imply that

$$\underset{\sim}{r}(s) = \int_0^s \{[1 + \xi(t)]\underset{\sim}{a}(t) + \eta(t)\underset{\sim}{b}(t)\}\, dt \equiv \underset{\sim}{r}[\underset{\sim}{u}](s), \tag{2.11}$$

$$\theta(s) = \int_0^s \mu(t)dt \equiv \theta[\underset{\sim}{u}](s). \tag{2.12}$$

Let $\underset{\sim}{n}(s)$ and $\underset{\sim}{m}(s)$ denote the resultant force and resultant couple applied to $[0,s]$ by $(s,1]$. Let the rod be subject to an external force of intensity $\underset{\sim}{f}(s)$ at s per unit reference length. We assume that $\underset{\sim}{f} \in C^{\circ}([0,1])$. If the only couples applied to the rod are at its ends, then the classical equilibrium equations are

$$\underset{\sim}{n}' + \underset{\sim}{f} = \underset{\sim}{0}, \qquad \underset{\sim}{m}' + \underset{\sim}{r}' \times \underset{\sim}{n} = \underset{\sim}{0}. \tag{2.13a,b}$$

We assume that the end $s = 1$ is free so that

$$\underset{\sim}{n}(1) = \underset{\sim}{0}, \qquad \underset{\sim}{m}(1) = \underset{\sim}{0}. \tag{2.14a,b}$$

(The reactions at $s = 1$ are found by integrating (2.13) over $[0,1]$.)

In consonance with our assumption of planarity, we decompose $\underset{\sim}{n}$ and $\underset{\sim}{m}$ thus:

$$\underset{\sim}{n}(s) = \Xi(s)\underset{\sim}{a}(s) + H(s)\underset{\sim}{b}(s), \qquad \underset{\sim}{m}(s) = M(s)\underset{\sim}{k} \tag{2.15a,b}$$

so that (2.13) reduces to

$$\Xi' - H\theta' + \underset{\sim}{f}\cdot\underset{\sim}{a} = 0, \tag{2.16a}$$

$$H' + \Xi\theta' + \underset{\sim}{f}\cdot\underset{\sim}{b} = 0, \tag{2.16b}$$

$$M' + (1 + \xi)H - y\Xi = 0. \tag{2.16c}$$

We set

$$\underset{\sim}{\sigma} = (\Xi,H,M). \tag{2.17}$$

We introduce the constitutive function

$$(\underset{\sim}{u},s) \in K \times [0,1] \longmapsto \underset{\sim}{\sigma}(\underset{\sim}{u},s) \in \mathbb{R}^3 \tag{2.18}$$

with $\hat{\sigma} \in [C^1(K \times [0,1])]^3$. The constitutive equations for an elastic rod whose configuration is given by (2.1) are

$$\sigma(s) = \hat{\sigma}(u(s),s). \tag{2.19}$$

Denoting partial derivatives by subscripts, we require that

$$\frac{\partial \hat{\sigma}}{\partial u} \equiv \begin{pmatrix} \hat{\Xi}_\xi & \hat{\Xi}_\eta & \hat{\Xi}_\mu \\ \hat{H}_\xi & \hat{H}_\eta & \hat{H}_\mu \\ \hat{M}_\xi & \hat{M}_\eta & \hat{M}_\mu \end{pmatrix} \tag{2.20}$$

be positive-definite, which implies that $u \longmapsto \hat{\sigma}(u,s)$ is a strictly monotone mapping. This is a very natural assumption (cf. Antman [7].) We impose the coercivity condition that

$$|\hat{\sigma}(u,s)| \longrightarrow \infty \quad \text{as} \quad |u| \longrightarrow \infty \text{ or as } \delta(u) \longrightarrow 0. \tag{2.21}$$

This condition will be strengthened slightly to handle various questions of analysis.

The rod is called hyperelastic if there exists a function

$$(u,s) \in K \times [0,1] \longrightarrow \psi(u,s) \in \mathbb{R} \tag{2.22}$$

such that

$$\hat{\sigma} = \partial\psi/\partial u. \tag{2.23}$$

We take $\psi \in C^2(K \times [0,1])$. The positive-definiteness of (2.20) implies that $\psi(\cdot,s)$ is strictly convex. For hyperelastic rods, we strengthen (2.22) by requiring that

$$\psi(\underset{\sim}{u},s) \longrightarrow \infty \quad \text{as} \quad |\underset{\sim}{u}| \longrightarrow \infty \quad \text{or as} \quad \delta(\underset{\sim}{u}) \longrightarrow 0. \tag{2.24}$$

It is convenient to define

$$\psi(\underset{\sim}{u},s) = \infty \quad \text{for} \quad \underset{\sim}{u} \notin K. \tag{2.25}$$

Then for each $s \in [0,1]$, $\psi(\cdot,s)$ is a lower semicontinuous convex function on $\mathbb{R}^3$, i.e., for each $a \in \mathbb{R}$

$$\{\underset{\sim}{u} \in \mathbb{R}^3 : \psi(\underset{\sim}{u},s) \leqslant a\} \quad \text{is a closed convex set.} \tag{2.26}$$

This means that

$$\psi(\underset{\sim}{u},s) \leqslant \liminf \psi(\underset{\sim}{u}_k,s) \quad \text{as} \quad \underset{\sim}{u}_k \longrightarrow \underset{\sim}{u} \quad \text{in} \quad \mathbb{R}^3. \tag{2.27}$$

Our boundary value problem consists of the strain configuration equations (2.11), (2.12), the equilibrium equations (2.16), the constitutive equations (2.20), the boundary conditions (2.14), and the strict unilateral constraint (2.8). Note that (2.11), (2.12) enable us to reduce this problem to one for $\underset{\sim}{u}$ alone.

Remarks The most general boundary value problem for ordinary differential equations of rod and shell theory are formulated by Antman [7]. In these problems (i) the configuration is defined by a function $s \in [0,1) \longmapsto \underset{\sim}{w}(s) \in \mathbb{R}^n$, (ii) orientation-preserving deformations are characterized by a relation of the form $(\underset{\sim}{w}(s),\underset{\sim}{w}'(s)) \in G(s)$ for all $s \in [0,1]$, where $G(s) \subset \mathbb{R}^{2n}$ has a number of important properties, prominent among which is the requirement that for each $\underset{\sim}{a} \in \mathbb{R}^n$, $\{(\underset{\sim}{w},\underset{\sim}{z}) \in G(s) : \underset{\sim}{w} = \underset{\sim}{a}\}$ is an open, convex, proper subset of $\mathbb{R}^n$, (iii) the classical equilibrium equations have the form $\underset{\sim}{\mu}' - \underset{\sim}{\nu} + \underset{\sim}{\phi} = \underset{\sim}{0}$, where the stress resultants $\underset{\sim}{\mu}$, $\underset{\sim}{\nu}$ map $[0,1]$ to $\mathbb{R}^n$ and the external load $\underset{\sim}{\phi}$ maps the space of configurations

$\times[0,1]$ to $\mathbb{R}_n$, (iv) the constitutive functions $(\underset{\sim}{w},\underset{\sim}{z}) \in G(s) \mapsto \hat{\underset{\sim}{\mu}}(\underset{\sim}{w},\underset{\sim}{z},s)$, $\hat{\underset{\sim}{\nu}}(w,z,s) \in \mathbb{R}^n$ meet the semi-monotonicity requirement that $\hat{\underset{\sim}{\mu}}(\underset{\sim}{w},\cdot,s)$ be strictly monotone.

One important feature of one-dimensional problems that does not occur in higher-dimensional problems is that the configurations can be expressed directly in terms of the strains as in (2.11), (2.12). In higher-dimensional problems, the possibility of obtaining such representations devolves upon the solvability of the Riemann-Christoffel compatibility equations. The high order of these nonlinear equations makes their analysis difficult, if not impossible, in the function spaces found most useful for studying corresponding boundary value problems for nonlinear elliptic systems.

3 DIRECT METHODS OF THE CALCULUS OF VARIATIONS FOR CONSERVATIVE PROBLEMS

Set

$$U_0[\underset{\sim}{u}] = \int_0^1 \psi(\underset{\sim}{u}(s),s)ds, \tag{3.1}$$

$$U_1[\underset{\sim}{u}] = \int_0^1 \underset{\sim}{f}(s)\cdot\underset{\sim}{r}[\underset{\sim}{u}](s)ds, \tag{3.2}$$

$$U = U_0 + U_1. \tag{3.3}$$

We study

3.4 Problem Minimize U over the admissible class A consisting of all Lebesgue measurable $\underset{\sim}{u}$ for which $U[\underset{\sim}{u}] < \infty$. Show that its minimizer is orientation-preserving and is a classical solution of the Euler equations and natural boundary conditions for U. (It is easy to show that these equations and boundary conditions are exactly those of our boundary value problem. Note that A is the largest class for which this problem makes sense.)

We shall strengthen the hypotheses on ψ so that we can prove that U is minimized on A by verifying the hypotheses of the abstract

<u>3.5 Minimization Theorem</u> A sequentially weakly lower semicontinuous functional on a non-empty, bounded, sequentially weakly closed subset of a reflexive Banach space is bounded below and attains its minimum there. The simple proof of this theorem is given by Vainberg [21], inter alia.

Since $\psi(\cdot,s)$ is lower semi-continuous on $\mathbb{R}^3$, the Minimization Theorem implies it is bounded below. Since our problem is unchanged by the addition of a constant to ψ, there is no loss of generality in taking ψ non-negative. We assume that there are constants $\alpha > 1$, $k_o > 0$, $k_1 > 0$ such that

$$\psi(\underset{\sim}{u},s) \geqslant k_o(|\xi|^\alpha + |\eta|^\alpha + |\mu|^\alpha) - k_1. \tag{3.6}$$

This inequality and the Hölder inequality imply that

$$\underset{\sim}{u} \in A \text{ if and only if } U_o[\underset{\sim}{u}] < \infty, \quad A \subset [L_\alpha(0,1)]^3 . \tag{3.7}$$

We denote the norm on $[L_\alpha(0,1)]^3$ by $\|\cdot\|$.

<u>3.8 Proposition</u> Let $\psi(\cdot,s)$ be convex and lower semi-continuous on $\mathbb{R}^3$ for almost all $s \in [0,1]$, let $\psi(\underset{\sim}{u},\cdot)$ be measurable for all $\underset{\sim}{u} \in \mathbb{R}^3$, and let ψ satisfy (3.6). Then for each $c \in \mathbb{R}$, the set

$$E(c) \equiv \{\underset{\sim}{u} : U_o[\underset{\sim}{u}] \leqslant c\} \tag{3.9}$$

is closed in the weak topology of $[L_\alpha(0,1)]^3$.

Remarks The hypotheses on ψ in this proposition are weaker than those we have imposed on ψ in Section 2. This proposition implies that U_o is sequentially weakly lower semi-continuous on $E(c)$ for all $c \in \mathbb{R}$. Note that A is not weakly closed: Let $\underset{\sim}{u}_k = (\xi_k,0,0)$ where $\{\xi_k\}$ is a sequence of constants converging to -1. Then $\underset{\sim}{u}_k \in A$ and $\underset{\sim}{u}_k$ converges uniformly to $\underset{\sim}{w} = (-1,0,0)$, which is not in A because $U_o[\underset{\sim}{w}] = \infty$.

Proof of Proposition 3.8 The convexity of $\psi(\cdot,s)$ implies that $E(c)$ is convex. Since a strongly closed convex set is weakly closed, we need only show that $E(c)$ is strongly closed. Accordingly, let $\underset{\sim}{u}_k \in E(c)$ and let $\underset{\sim}{u}_k \longrightarrow \underset{\sim}{w}$ in $[L_\alpha(0,1]^3$. Then $\{\underset{\sim}{u}_k\}$ has a subsequence, which we also denote by $\{\underset{\sim}{u}_k\}$, that converges a.e. to $\underset{\sim}{w}$. The lower semi-continuity of $\psi(\cdot,s)$ and Fatou's Lemma imply that

$$U_o[\underset{\sim}{w}] \equiv \int_0^1 \psi(\underset{\sim}{w}(s),s)ds \leqslant \int_0^1 \liminf \psi(\underset{\sim}{u}_k(s),s)ds$$

$$\leqslant \liminf \int_0^1 \psi(\underset{\sim}{u}_k(s),s)ds \leqslant c. \qquad \square \tag{3.10}$$

3.11 Proposition U_1 is sequentially weakly continuous on $[L_\alpha(0,1)]^3$.

Proof. The use of Hölder's inequality and the Arzelà-Ascoli Theorem shows that $\underset{\sim}{u} \in [L_\alpha(0,1)]^3 \longmapsto \theta[\underset{\sim}{u}] \in C^\circ([0,1])$ is compact, so that the weak convergence of $\underset{\sim}{u}_k$ implies the strong convergence of $\theta[\underset{\sim}{u}_k]$, $\cos\theta[\underset{\sim}{u}_k]$, $\sin\theta[\underset{\sim}{u}_k]$. The weak continuity of U_1 follows by a direct appeal to the definition. $\square$

3.12 Theorem The hypotheses of Proposition 3.8 imply that U is minimized on A.

Proof. A is clearly non-empty. Inequality (3.6) implies that $U_0[\underset{\sim}{u}] \longrightarrow \infty$ as $\|\underset{\sim}{u}\| \longrightarrow \infty$ while (3.6) and Hölder's inequality show that $U[\underset{\sim}{u}] \longrightarrow \infty$ as $\|\underset{\sim}{u}\| \longrightarrow \infty$. It follows that there is a number C such that the infimum of U on A cannot be found on $\{\underset{\sim}{u}:U[\underset{\sim}{u}] > C\}$ and that there is a number c such that

$$\emptyset \neq \{\underset{\sim}{u}:U[\underset{\sim}{u}] \leqslant C\} \subset E(c) \subset A. \tag{3.13}$$

Since, by our preceeding remarks, $E(c)$ is bounded and is weakly closed by Proposition 3.8 and since U is weakly lower semi-continuous on $E(c)$ by Proposition 3.8 and 3.11, the conclusion follows from Theorem 3.5. □

We now turn to the deeper problem of showing that the minimizer of Theorem 3.12 is a classical solution of our boundary value problem and, in particular, satisfies (2.8) everywhere. We describe three different approaches with different advantages that enable us to resolve this regularity question. We compare these approaches at the end of this section.

Method I

A standard argument using the Lebesgue Dominated Convergence Theorem shows that U_1 is Gâteaux differentiable on $[L_\alpha(0,1)]^3$. Denote its Gâteaux differential at $\underset{\sim}{w}$ in the direction $\underset{\sim}{v}$ by

$$U_1'[\underset{\sim}{w}]\underset{\sim}{v} \equiv \int_0^1 \underset{\sim}{f}(s)\cdot\underset{\sim}{r}'[\underset{\sim}{w}](s)\cdot\underset{\sim}{v}(s)ds \equiv \int_0^1 \underset{\sim}{\rho}[\underset{\sim}{w}](s)\cdot\underset{\sim}{v}(s)ds. \tag{3.14}$$

Our basic result is given in the following theorem, in which we relax a number of our restrictions on ψ.

3.15 Theorem Let $\psi(\cdot,s)$ be convex on $\mathbb{R}^3$ for almost all $s \in [0,1]$, let there exist a function $\underset{\sim}{z}$ such that $\psi(\underset{\sim}{z}(\cdot),\cdot) \in L_\infty(0,1)$, and let U_1 be Gâteaux differentiable on $[L_\alpha(0,1)]^3$ with differential given by (3.14).

Let $\underset{\sim}{w}$ be a local minimizer of U on $[L_\alpha(0,1)]^3$. Then there exists a number C such that

$$\psi(\underset{\sim}{w}(s),s) \leqslant \underset{\sim}{\rho}[\underset{\sim}{w}](s)\cdot[\underset{\sim}{z}(s) - \underset{\sim}{w}(s)] + C, \quad \text{a.e.} \tag{3.16}$$

<u>Proof</u>. Since $\underset{\sim}{w}$ is a local minimizer, there is a number $\varepsilon > 0$ such that

$$\int_0^1 \psi(\underset{\sim}{w}(s),s)ds + U_1[\underset{\sim}{w}] < \int_0^1 \psi(\underset{\sim}{v}(s),s)ds + U_1[\underset{\sim}{v}] \tag{3.17}$$

for all $\underset{\sim}{v}$ satisfying $\|\underset{\sim}{v} - \underset{\sim}{w}\| \leqslant \varepsilon$. Let $\underset{\sim}{u}$ satisfy $\|\underset{\sim}{u} - \underset{\sim}{w}\| \leqslant \varepsilon$. Choose $\underset{\sim}{v} = (1 - t)\underset{\sim}{w} + t\underset{\sim}{u}$ for $t \in (0,1)$. Then the convexity of $\psi(\cdot,s)$ implies that

$$\int_0^1 \psi(\underset{\sim}{w}(s),s)ds \leqslant \int_0^1 \psi(\underset{\sim}{u}(s),s)ds + t^{-1}\{U_1[(1 - t)\underset{\sim}{w} + t\underset{\sim}{u}] - U_1[\underset{\sim}{w}]\} \tag{3.18}$$

for all $\underset{\sim}{u}$ satisfying $\|\underset{\sim}{u} - \underset{\sim}{w}\| \leqslant \varepsilon$. Let $t \longrightarrow 0$. Then (3.14) implies that

$$\int_0^1 \{\psi(\underset{\sim}{w}(s),s) - \psi(\underset{\sim}{u}(s),s) + \underset{\sim}{\rho}[\underset{\sim}{w}](s)\cdot[\underset{\sim}{w}(s) - \underset{\sim}{u}(s)]\}ds \leqslant 0 \tag{3.19}$$

for all such $\underset{\sim}{u}$. Now let

$$\underset{\sim}{u}(s) = \begin{cases} \underset{\sim}{z}(s) & \text{for } s \in \Omega, \\ \underset{\sim}{w}(s) & \text{for } s \notin \Omega, \end{cases} \tag{3.20}$$

where Ω is an arbitrary measurable subset of $[0,1]$ with Lebesgue measure $|\Omega|$ so small that $\|\chi_\Omega(\underset{\sim}{u} - \underset{\sim}{z})\| \leqslant \varepsilon$. χ_Ω is the characteristic function for Ω. Then (3.19) implies that

$$\int_\Omega \{\psi(\underset{\sim}{w}(s),s) - \psi(\underset{\sim}{z}(s),s) + \underset{\sim}{\rho}[\underset{\sim}{w}](s)\cdot[\underset{\sim}{w}(s) - \underset{\sim}{z}(s)]\}ds \leqslant 0 \tag{3.21}$$

for all such Ω, whence it follows that

$$\psi(\underset{\sim}{w}(s),s) \leqslant \text{ess sup } \psi(\underset{\sim}{z}(s),s) + \underset{\sim}{\rho}[\underset{\sim}{w}](s)\cdot[\underset{\sim}{z}(s) - \underset{\sim}{w}(s)], \quad \text{a.e. } \square. \tag{3.22}$$

To complete our regularity theory, we note that (3.16) implies that $\psi(\underset{\sim}{w}(\cdot),\cdot) \in L_\infty(0,1)$ for our problem. In particular, this means that $\underset{\sim}{w}(s)$ is essentially bounded and essentially bounded away from ∂K. We can now use the Lebesgue Dominated Convergence Theorem to show that U_0 has a Gâteaux differential at $\underset{\sim}{w}$ in the direction $\underset{\sim}{v}$ for each $\underset{\sim}{v} \in [L_\alpha(0,1)]^3$. Thus the Gâteaux derivative of U vanishes at $\underset{\sim}{w}$, i.e., $\underset{\sim}{w}$ is a weak solution of the Euler equations for U. The rest of the regularity theory is standard (cf., Akhiezer [1]).

Method II

We introduce the new variable

$$\zeta(\underset{\sim}{u}) \equiv \left[(1 + \xi)^2 - h^2\mu^2\right]^{\frac{1}{2}}, \tag{3.23}$$

which is positive for $\underset{\sim}{u} \in K$. ζ is continuously differentiable on K, a virtue not enjoyed by δ. We strengthen the hypotheses on ψ by assuming that it has the form

$$\psi(\underset{\sim}{u},s) = p(\underset{\sim}{u},s) + q(\underset{\sim}{u},s), \quad p(\underset{\sim}{u},s) = P(\zeta(\underset{\sim}{u}),s) \tag{3.24}$$

where

$$(\zeta,s) \in (0,\infty) \times [0,1] \longmapsto P(\zeta,s) \in [0,\infty)$$

is a twice continuously differentiable function satisfying

$$\frac{\partial P}{\partial \zeta}(\zeta,s) \leqslant 0, \quad \frac{\partial^2 P}{\partial \zeta^2}(\zeta,s) \geqslant 0, \quad P(\zeta,s) \longrightarrow \infty \text{ as } \zeta \searrow 0, \tag{3.25a,b,c}$$

and where

$$(\underset{\sim}{u},s) \in \mathbb{R}^3 \times [0,1] \longmapsto q(\underset{\sim}{u},s) \in [0,\infty)$$

is a twice continuously differentiable function with the property that there are continuous functions $\alpha_1,\alpha_2,\alpha_3:[0,1] \longrightarrow (1,\infty)$ and that there are positive numbers k_0,k_1,k_2 such that

$$q(\underset{\sim}{u},s) \geqslant k_0\left[|\xi|^{\alpha_1(s)} + |\eta|^{\alpha_2(s)} + |\mu|^{\alpha_3(s)}\right] - k_1, \tag{3.26}$$

$$\left|\frac{\partial q}{\partial \underset{\sim}{u}}(\underset{\sim}{u},s)\right| < k_2\left[|\xi|^{\alpha_1(s)} + |\eta|^{\alpha_2(s)} + |\mu|^{\alpha_3(s)} + 1\right]. \tag{3.27}$$

The assumptions (3.24)-(3.27) are compatible with the assumptions on ψ made in Section 2. In particular, if $P(\zeta,s)$ is taken to be equal to $\beta(s)\zeta^{-\sigma}$, where $\beta(s) > 0$ and $\sigma > 0$, then $P(\cdot,s)$ is convex. We could readily generalize (3.24)-(3.27) by relaxing (3.25) and by letting P also depend upon η and μ. (No further generality is gained by allowing P to depend on (ζ,ξ,η,μ,s) rather than just on (ζ,η,μ,s) because (3.23) induces a bijection from $\underset{\sim}{u} \in K$ to $(\zeta,\eta,\mu) \in (0,\infty)\times\mathbb{R}^2$.) We do not pursue such generalizations because they obscure the principal techniques for treating orientation-preserving deformations.

Inequality (3.26) implies that A is contained in

$$L \equiv \left\{\underset{\sim}{u}: \int_0^1 \left[|\xi(s)|^{\alpha_1(s)} + |\eta(s)|^{\alpha_2(s)} + |\mu(s)|^{\alpha_3(s)}\right] ds < \infty\right\} \tag{3.28}$$

L can be made into a reflexive Banach space (cf., Krasnosel'skii & Rutitskii [16]). We denote that norm of L by $\|\cdot\|$. Note that there is an $\alpha > 1$ such that $L \subset [L_\alpha(0,1)]^3$. Thus (3.6) and its consequences, namely Theorem 3.12, hold.

3.29 Theorem Let $\underset{\sim}{w} = (\xi,\eta,\mu)$ be a local minimizer of U on A. Let ψ satisfy (3.24)-(3.27). Then $\underset{\sim}{w}$ is a classical solution of the boundary value problem and satisfies (2.8) everywhere.

Proof. Let

$$\Omega_\varepsilon = \{s \in [0,1] : \zeta(\underset{\sim}{w}(s)) > \varepsilon\} \quad \text{for } 0 \leqslant \varepsilon < \infty, \quad \Omega^C_\varepsilon = [0,1]\setminus\Omega_\varepsilon. \tag{3.30}$$

Let $\dot{\underset{\sim}{u}} = (\dot{\xi},\dot{\eta},\dot{\mu}) \in [L_\infty(0,1)]^3$ with $\dot{\underset{\sim}{u}} = 0$ on Ω^C_ε. Since $\underset{\sim}{w}$ is a local minimizer of U, we have

$$\begin{aligned} 0 &\leqslant t^{-1}\{U[\underset{\sim}{w} + t\dot{\underset{\sim}{u}}] - U[\underset{\sim}{w}] \\ &= \int_{\Omega_\varepsilon} t^{-1}[p(\underset{\sim}{w}(s) + t\dot{\underset{\sim}{u}}(s),s) - p(\underset{\sim}{w}(s),s)]ds \\ &+ \int_{\Omega_\varepsilon} t^{-1}[q(\underset{\sim}{w}(s) + t\dot{\underset{\sim}{u}}(s),s) - q(\underset{\sim}{w}(s),s)]ds \\ &+ \int_{\Omega_\varepsilon} t^{-1}\underset{\sim}{f}(s)\cdot\{\underset{\sim}{r}[\underset{\sim}{w} + t\dot{\underset{\sim}{u}}](s) - \underset{\sim}{r}[\underset{\sim}{w}](s)\}ds \end{aligned} \tag{3.31}$$

for sufficiently small positive t. Since ζ is continuous on K and since $\dot{\underset{\sim}{u}} \in [L_\infty(0,1)]^3$, there is a number $t_\varepsilon > 0$ such that $\zeta(\underset{\sim}{w}(s) + t\dot{\underset{\sim}{u}}(s)) \geqslant \varepsilon/2$ for $s \in \Omega_\varepsilon$, $t \in [0,t_\varepsilon]$. The mean value theorem and (3.25a,b) imply that

$$\begin{aligned} &t^{-1}|p(\underset{\sim}{w}(s) + t\dot{\underset{\sim}{u}}(s),s) - p(\underset{\sim}{w}(s),s)| \\ &\leqslant \frac{2}{\varepsilon}\left|\frac{\partial P}{\partial\zeta}\left(\frac{\varepsilon}{2},s\right)\right|\cdot\left[(1 + \xi(s) + t_\varepsilon|\dot{\xi}(s)|)|\dot{\xi}(s)| + h^2(|\mu(s)| + t_\varepsilon|\dot{\mu}(s)|)|\dot{\mu}(s)|\right] \end{aligned} \tag{3.32}$$

for $t \in (0,t_\varepsilon]$, $s \in \Omega_\varepsilon$. The right side of (3.32) is the value of an integrable function on Ω_ε. The mean value theorem and (3.28) imply that the absolute value of the integrand of the second integral of (3.31) is dominated by an integrable function. The same conclusion is readily demonstrated for the third integral. Since the integrands of these integrals converge pointwise to their obvious limits, the Lebesgue Dominated Convergence Theorem implies that the limit of (3.31) as $t \searrow 0$ is given by

$$0 \leqslant \int_{\Omega_\varepsilon} \left\{ \frac{\partial\psi}{\partial \underset{\sim}{u}}(\underset{\sim}{w}(s),s) + \underset{\sim}{\rho}[\underset{\sim}{w}](s) \right\} \cdot \dot{\underset{\sim}{u}}(s)\,ds \qquad \forall\, \dot{\underset{\sim}{u}} \in [L_\infty(\Omega_\varepsilon)]^3. \tag{3.33}$$

The arbitrariness of $\dot{\underset{\sim}{u}}$ implies that

$$\frac{\partial\psi}{\partial \underset{\sim}{u}}(\underset{\sim}{w}(s),s) = -\underset{\sim}{\rho}[\underset{\sim}{w}](s) \tag{3.34}$$

for almost all s in Ω_ε. The arbitrariness of ε implies that (3.34) holds for almost all s in $\bigcup_{\varepsilon>0} \Omega_\varepsilon = \Omega_0$. But (3.26c) and the inequality $U[\underset{\sim}{w}] > \infty$ imply that $|\Omega_0| = 1$ so (3.34) holds a.e. Since the right side of (3.34) is continuous, the left is also. Thus (3.24) and (3.26c) imply that $\zeta(\underset{\sim}{w}(s))$ is essentially bounded away from 0. The rest of the proof is standard. □

There are several variants of this approach. We describe a somewhat more cumbersome process, which is better suited for handling more complicated problems like those described at the end of Section 2.

We chose $\dot{\underset{\sim}{u}}$ such that

$$\dot{\xi}(s) \geqslant 0, \quad \text{sign}\ \dot{\mu}(s) = -\text{sign}\ \mu(s) \quad \text{for}\ s \in [0,1]. \tag{3.35}$$

Then (3.31) is replaced by

$$0 \leqslant \liminf_{\varepsilon \searrow 0} \ \limsup_{t \searrow 0} t^{-1}\{U[\underset{\sim}{w} + t\dot{\underset{\sim}{u}}] - U[\underset{\sim}{w}]\} \tag{3.36a}$$

$$\leqslant \liminf_{\varepsilon \searrow 0} \ \limsup_{t \searrow 0} t^{-1}\int_0^1 \Big\{\chi_{\Omega_\varepsilon}[p(\underset{\sim}{w} + t\dot{\underset{\sim}{u}},s) - p(\underset{\sim}{w},s)]$$

$$+ [q(\underset{\sim}{w} + t\dot{\underset{\sim}{u}},s) - q(\underset{\sim}{w},s)] + \underset{\sim}{f}\cdot(\underset{\sim}{r}[\underset{\sim}{w} + t\dot{\underset{\sim}{u}}] - \underset{\sim}{r}[\underset{\sim}{w}])\Big\}ds \tag{3.36b}$$

$$= \liminf_{\varepsilon \searrow 0} \int_0^1 \Big\{\chi_{\Omega_\varepsilon} \frac{\partial p}{\partial \underset{\sim}{u}}(\underset{\sim}{w},s) + \frac{\partial q}{\partial \underset{\sim}{u}}(\underset{\sim}{w},s) + \underset{\sim}{\rho}[\underset{\sim}{w}]\Big\}\cdot\dot{\underset{\sim}{u}}\, ds \tag{3.36c}$$

$$\leqslant \int_0^1 \Big\{\frac{\partial p}{\partial \underset{\sim}{u}}(\underset{\sim}{w},s) + \frac{\partial q}{\partial \underset{\sim}{u}}(\underset{\sim}{w},s) + \underset{\sim}{\rho}[\underset{\sim}{w}]\Big\}\cdot\dot{\underset{\sim}{u}}\, ds \tag{3.36d}$$

for all such $\dot{\underset{\sim}{u}} \in [L_\infty(0,1)]^3$. Inequality (3.36a) is justified by the local minimization of U at $\underset{\sim}{w}$. Inequality (3.36b) follows from the non-negativity of

$$\int_{\Omega_\varepsilon^c} [p(\underset{\sim}{w} + t\dot{\underset{\sim}{u}},s) - p(\underset{\sim}{w},s)]ds\,, \tag{3.37}$$

which is a consequence of (3.26a) and (3.35). Equation (3.36c) follows from the Lebesgue Dominated Convergence Theorem. Now (3.25a) and (3.35) also imply that the integrand of (3.36c) is bounded above by the integrable function $s \longmapsto \frac{\partial q}{\partial \underset{\sim}{u}}(\underset{\sim}{w}(s),s)\cdot\dot{\underset{\sim}{u}}(s) + \underset{\sim}{\rho}[\underset{\sim}{w}](s)\cdot\dot{\underset{\sim}{u}}(s)$. Relation (3.25c) and the inequality $U[\underset{\sim}{w}] < \infty$ imply that $\chi_{\Omega_\varepsilon}(s) \longrightarrow 0$ a.e. as $\varepsilon \longrightarrow 0$. These facts enable us to use a version of Fatou's lemma (cf. Royden [19], Prob. 4.13) to justify inequality (3.36d). We finally conclude from the entire chain of inequalities of (3.36) and from (3.25a), (3.35) that $\frac{\partial p}{\partial \underset{\sim}{u}}(\underset{\sim}{w}(\cdot),\cdot)$ is integrable on $[0,1]$. We now suspend (3.35), letting $\dot{\underset{\sim}{u}}$ vary over all $L_\infty(0,1)^3$. The integrability of $\frac{\partial \psi}{\partial \underset{\sim}{u}}(\underset{\sim}{w}(\cdot),\cdot)$ enables us to apply Lebesgue's Dominated Convergence Theorem to (3.33) for all $\underset{\sim}{u} \in [L_\infty(0,1]^3$, whence we obtain

$$0 = \int_0^1 \left\{ \frac{\partial \psi}{\partial \underset{\sim}{u}} (\underset{\sim}{w}(s),s) + \underset{\sim}{\rho}[\underset{\sim}{w}](s) \right\} \cdot \dot{\underset{\sim}{u}}(s)ds \qquad \forall\ \dot{\underset{\sim}{u}} \in [L_\infty(0,1)]^3, \tag{3.38}$$

i.e., $\underset{\sim}{w}$ is a weak solution of the boundary value problem. The application of a version of the Fundamental Lemma of the Calculus of Variations immediately yields (3.34) a.e.

Method III

Let g be a twice continuously differentiable, strictly increasing function $\mathbb{R}$ onto $(0,\infty)$ for which there exists a number $G > 0$ such that

$$|g(\omega)| \leqslant G(1 + |\omega|). \tag{3.39}$$

(This inequality can be relaxed.) We define a new strain variable ω by

$$\omega = g^{-1}(\zeta(\underset{\sim}{u})). \tag{3.40}$$

ω has the virtue of ranging over $\mathbb{R}$ as $\zeta(\underset{\sim}{u})$ ranges over $(0,\infty)$. Moreover $\underset{\sim}{u} \in K$ if and only if $\omega > -\infty$. We introduce a new set of unconstrained strains

$$\underset{\sim}{v} = (\omega,\eta,\mu). \tag{3.41}$$

From (3.23) and (3.40), we have

$$1 + \xi = \left[g(\omega)^2 + h^2\mu^2\right]^{\frac{1}{2}}. \tag{3.42}$$

We define

$$\underset{\sim}{\gamma}(\underset{\sim}{v}) \equiv \left(\left[g(\omega)^2 + h^2\mu^2\right]^{\frac{1}{2}} - 1,\eta,\mu\right) \tag{3.43}$$

so that $\underset{\sim}{u} = \underset{\sim}{\gamma}(\underset{\sim}{v})$. $\underset{\sim}{\gamma}$ is a bijection from $\mathbb{R}^3$ to K that takes measurable functions into measurable function (cf. Hewitt & Stromberg [13], Thm. 11.7).

We set

$$V_k[\underset{\sim}{v}] = U_k[\underset{\sim}{\gamma}(\underset{\sim}{v}(\cdot))], \quad k = 0,1, \quad V[\underset{\sim}{v}] = U[\underset{\sim}{\gamma}(\underset{\sim}{v}(\cdot))], \tag{3.44}$$

$$\phi(\underset{\sim}{v},s) = \psi(\underset{\sim}{\gamma}(\underset{\sim}{v}),s). \tag{3.45}$$

<u>3.46 Theorem</u> Let U be minimized on A. Let there be continuous functions $\alpha_1,\alpha_2,\alpha_3:[0,1] \longrightarrow (1,\infty)$ and numbers $k_0,k_1,k_2 > 0$ such that

$$\phi(\underset{\sim}{v},s) \geqslant k_0\left(|\omega|^{\alpha_1(s)} + |\eta|^{\alpha_2(s)} + |\mu|^{\alpha_3(s)}\right) - k_1, \tag{3.47}$$

$$\frac{\partial\phi}{\partial \underset{\sim}{v}}(\underset{\sim}{v},s) \leqslant k_2\left(|\omega|^{\alpha_1(s)} + |\eta|^{\alpha_2(s)} + |\mu|^{\alpha_3(s)} + 1\right). \tag{3.48}$$

(We postulate neither (3.26) nor (3.27), so the functions $\alpha_1,\alpha_2,\alpha_3$ and the numbers k_0,k_1,k_2 appearing in (3.47),(3.48) need bear no relation to those of (3.26) or (3.27).) Then U is minimized on A and its minimizer is a classical solution of the boundary value problem and satisfies (2.8) everywhere.

<u>Proof.</u> Since U is minimized on A, V is minimized on the class of measurable $\underset{\sim}{v}$'s for which $V[\underset{\sim}{v}] < \infty$. Call the minimizer $\underset{\sim}{z}$. Now $V_0[\underset{\sim}{v}] \longrightarrow \infty$ as $\|\underset{\sim}{v}\| \longrightarrow \infty$ by (3.47) and $V[\underset{\sim}{v}] \longrightarrow \infty$ as $\|\underset{\sim}{v}\| \longrightarrow \infty$ by this result and by (3.39). Thus $\underset{\sim}{z} \in L$. Relation (3.48) enables us to use the Lebesgue Dominated Convergence Theorem to show that V is Gâteaux differentiable on L. This Gâteaux derivative vanishes at $\underset{\sim}{z}$. The classical methods of the Calculus of Variations show that $\underset{\sim}{z}$ is actually a classical solution of the Euler equations for V. Since $\underset{\sim}{z}$ has a bounded range, the minimizer $\underset{\sim}{w} = \underset{\sim}{\gamma}(\underset{\sim}{z}(\cdot))$ of U satisfies (2.8) everywhere by the properties of $\underset{\sim}{\gamma}$. □

Discussion of the Various Methods

Method I, which has the weakest assumptions, is especially simple and elegant. The results do not depend on the dimension of the region in which the independent variables lie. The proof of Theorem 3.15 relies strongly on the convexity of $\psi(\cdot,s)$. It seems possible to extend this theorem to certain semi-convex functionals like those described at the end of Section 2 by using methods of Antman [8]; it is not evident, however, that this method can handle three-dimensional hyperelasticity when the energy function merely satisfies the strong Legendre-Hadamard condition or Ball's [10] restriction thereof. The use of Method I in elasticity is new.

Method II employs somewhat sharper restrictions on ψ than Method I. Many of these restrictions can be considerable relaxed, however. Method II has the major advantage that it does not depend so strongly on convexity and the minor advantage that it delivers the weak form of equations directly. Method II is developed for the general problem described at the end of Section 2 by Antman [8].

Method III has the advantage that it avoids the measure theoretic arguments of Methods I and II. It has the minor disadvantage that the $\underset{\sim}{v}$-problem is somewhat artificial and the major disadvantage that it relies in an essential way on the representations (2.11), (2.12) giving the configuration in terms of the strain. Similar transformed strains can be introduced into the three-dimensional theory by replacing the right stretch tensor $\underset{\sim}{U}$ by $g^{-1}(\underset{\sim}{U})$, where g has the properties described in the paragraph containing (3.39), (3.40). But the utility of $g^{-1}(\underset{\sim}{U})$ in the three-dimensional theory is greatly reduced by the compatibility problem of representing position in terms of it. See the discussion at the end of Section 2. Method III was developed by Antman [3], [4], [5].

We finally mention another device that is very useful for the more complicated problems described at the end of Section 2 (cf. Antman [8]): If y is Lipschitz continuous from $[0.1]$ to $\mathbb{R}$ and if $\int_0^1 [y(s)]^{-1} ds < \infty$, then y cannot vanish on $[0,1]$ and is accordingly uniformly bounded away from 0 on $[0,1]$.

Antman's [8] treatment of the problems described at the end of Section 2 relied strongly on the one-dimensional character of the problem in its use of the facts that (i) elements of Sobolev spaces on the line are Hölder continuous and (ii) open subsets of the line are countable unions of open intervals. (Property (i) is used implicitly in the proof of Proposition 3.18. These results are not available for problems in higher dimensions.)

For other choices of U_1, such as that corresponding to a hydrostatic pressure $U[\underset{\sim}{u}]$ need not approach ∞ as $\|\underset{\sim}{u}\| \longrightarrow \infty$. In such cases the proof of Theorem 3.12 breaks down for the good reason that the theorem need not be true. There are, however, related variational problems that give existence proofs for the boundary value problem (cf. Antman [4], [5], [8]).

4 MONOTONE OPERATOR METHODS FOR NONCONSERVATIVE PROBLEMS

In Section 3, we faced the problem that $\frac{\partial\psi}{\partial \underset{\sim}{u}}(\underset{\sim}{u}(\cdot),\cdot)$ need not belong to $[L_1(0,1)]^3$ even if $U_0[\underset{\sim}{u}] \leqslant c$. (We effectively showed, however, that $\frac{\partial\psi}{\partial \underset{\sim}{u}}(\underset{\sim}{w}(\cdot),\cdot) \in [L_1(0,1)]^3$, when $\underset{\sim}{w}$ is the minimizer of Theorem 3.12.) One might be tempted to circumvent this problem by assuming that there is a function $(\underset{\sim}{u},s) \longmapsto \Psi(\underset{\sim}{u},s)$ with $\Psi(\cdot,s)$ convex and lower semi-continuous such that

$$\psi(\underset{\sim}{u},s) + \left|\frac{\partial\psi}{\partial \underset{\sim}{u}}(\underset{\sim}{u},s)\right| \leqslant \Psi(\underset{\sim}{u},s) \tag{4.1}$$

and by seeking a minimum for U on $B = \left\{ \underset{\sim}{u} : \int_0^1 \Psi(\underset{\sim}{u},s)ds < \infty \right\}$ rather than on the larger set $A = \left\{ \underset{\sim}{u} : \int_0^1 \psi(\underset{\sim}{u},s)ds < \infty \right\}$. If U could be shown to have a minimum on B, then the regularity theory would be easy because (4.1) would imply that U is Gâteaux differentiable on B and therefore at $\underset{\sim}{w}$. Then (3.38) holds, whence it follows that (3.34) holds a.e. This means that $\underset{\sim}{w}$ is regular and satisfies (2.8) everywhere.

Although $E(c)$ is weakly closed, B is not, so there is no point in seeking a minimum for U on $E(c) \cap B$. Moreover, there are $\underset{\sim}{u}$'s in A for which $\int_0^1 \Psi(\underset{\sim}{u}(s),s)ds = \infty$, so there is no point in seeking a minimum for U on a weakly closed set of the form $\left\{ \underset{\sim}{u} : \int_0^1 \Psi(\underset{\sim}{u}(s),s)ds \leq c \right\}$. Thus (4.1) does not seem to present a feasible approach for handling the problem of minimizing U in the spirit of Theorem 3.5. It is therefore somewhat paradoxical that an analog of (4.1) nevertheless forms the basis for a successful treatment of boundary value problems that are not necessarily conservative. We turn to the study of these.

Motivated by (3.33), (3.36), which are based upon (2.23), we study the variational inequality of finding a strain with values in K such that

$$\langle \underset{\sim}{\sigma}(\underset{\sim}{w}(\cdot),\cdot) + \underset{\sim}{\rho}[\underset{\sim}{w}], \underset{\sim}{u} - \underset{\sim}{w} \rangle \equiv \int_0^1 \{ \underset{\sim}{\sigma}(\underset{\sim}{w}(s),s) + \underset{\sim}{\rho}[\underset{\sim}{w}](s)\} \cdot [\underset{\sim}{u}(s) - \underset{\sim}{w}(s)]ds \geq 0 \tag{4.2}$$

for all $\underset{\sim}{u}$ in a large class of functions with values in K for which the integral in (4.2) is defined.

We assume that there is a real valued function $\Psi \in C^1(K \times [0,1])$ such that

$\Psi(\cdot,s)$ is strictly convex for each $s \in [0,1]$, (4.3)

$\Psi(\underset{\sim}{u},s) \longrightarrow \infty$ as $|\underset{\sim}{u}| \longrightarrow \infty$ and as $\delta(\underset{\sim}{u}) \longrightarrow 0$, (4.4)

there are continuous functions $\alpha_1,\alpha_2,\alpha_3 : [0,1] \longrightarrow (1,\infty)$ and positive numbers k_0, k_1 such that

$$\Psi(\underset{\sim}{u},s) \geq k_0\left(|\xi|^{\alpha_1(s)} + |\eta|^{\alpha_2(s)} + |\mu|^{\alpha_3(s)}\right) - k_1, \tag{4.5}$$

$$\hat{\sigma}(\underset{\sim}{u},s)\cdot\frac{\partial\Psi(\underset{\sim}{u},s)/\partial\underset{\sim}{u}}{|\partial\Psi(\underset{\sim}{u},s)/\partial\underset{\sim}{u}|} \longrightarrow \infty \text{ as } \delta(\underset{\sim}{u}) \longrightarrow 0,$$

$$|\underset{\sim}{u}|^{-1}\,\hat{\sigma}(\underset{\sim}{u},s)\cdot\frac{\partial\Psi(\underset{\sim}{u},s)/\partial\underset{\sim}{u}}{|\partial\Psi(\underset{\sim}{u},s)/\partial\underset{\sim}{u}|} \longrightarrow \infty \text{ as } |\underset{\sim}{u}| \longrightarrow \infty. \tag{4.6}$$

$\hat{\sigma}(\underset{\sim}{u}(\cdot),\cdot) \in L^*$ (the dual space of L) when $\int_0^1 \Psi(\underset{\sim}{u}(s),s)ds < \infty$. (4.7)

A sufficient condition for (4.7) is that there exist functions $a \in L_\infty(0,1)$, $b \in L_1(0,1)$ such that

$$|\hat{\Xi}(\underset{\sim}{u},s)|^{\alpha_1^*} + |\hat{H}(\underset{\sim}{u},s)|^{\alpha_2^*} + |\hat{M}(\underset{\sim}{u},s)|^{\alpha_3^*} \leq a(s)\Psi(\underset{\sim}{u},s) + b(s) \tag{4.8}$$

where

$$\alpha_k^* = \alpha_k/(\alpha_k - 1).$$

<u>4.9 Theorem</u> Let Ψ have the properties (4.3)-(4.7). Then there is a $\underset{\sim}{w}$ in

$$B \equiv \left\{\underset{\sim}{u} : \int_0^1 \Psi(\underset{\sim}{u}(s),s)ds < \infty\right\} \tag{4.10}$$

that satisfies (4.2) for all $\underset{\sim}{u}$ in B and, indeed, satisfies

$$\hat{\sigma}(\underset{\sim}{w}(s),s) + \underset{\sim}{\rho}[\underset{\sim}{w}](s) = \underset{\sim}{0} \text{ a.e.} \tag{4.11}$$

<u>Proof</u>. Let

$$F(c) \equiv \left\{ \underset{\sim}{u} : \int_0^1 \Psi(\underset{\sim}{u}(s),s)ds \leqslant c \right\}. \tag{4.12}$$

The results of Section 3 show that $F(c)$ is a closed, convex, bounded subset of the reflexive Banach space L. The monotonicity of $\hat{\underset{\sim}{\sigma}}(\cdot,s)$ on K implies the monotonicity of $\underset{\sim}{u}(\cdot) \longmapsto \hat{\underset{\sim}{\sigma}}(\underset{\sim}{u}(\cdot),\cdot)$ on B and hence the pseudo-monotonicity of $\underset{\sim}{u}(\cdot) \longmapsto \hat{\underset{\sim}{\sigma}}(\underset{\sim}{u}(\cdot),\cdot) + \rho[u]$ on $F(c)$. (Cf. Lions [17]). Let n be a positive integer. A theorem of Brezis [11] (cf. Lions [17] Ch. 2.8) implies that there is an element $\underset{\sim}{w}_n \in F(n)$ such that

$$\langle \hat{\underset{\sim}{\sigma}}(\underset{\sim}{w}_n(\cdot),\cdot) + \underset{\sim}{\rho}[\underset{\sim}{w}_n], \underset{\sim}{u} - \underset{\sim}{w}_n \rangle \geqslant 0 \qquad \forall\, \underset{\sim}{u} \in F(n). \tag{4.13}$$

We now wish to exploit the arbitrariness of $\underset{\sim}{u}$ in (4.13) to replace $\underset{\sim}{u} - \underset{\sim}{w}_n$ with some scalar multiple of $\partial\Psi(\underset{\sim}{w}_n(\cdot),\cdot)/\partial\underset{\sim}{u}$ in order to take advantage of the coercivity condition (4.6). Let $\varepsilon > 0$. There is a function $\underset{\sim}{u}_\varepsilon$ such that

$$\underset{\sim}{u}_\varepsilon(s) + \varepsilon \frac{\partial\Psi}{\partial\underset{\sim}{u}} (\underset{\sim}{u}_\varepsilon(s),s) = \underset{\sim}{w}_n(s) \tag{4.14}$$

because $\underset{\sim}{u} \in \overline{K} \longrightarrow \frac{1}{2}\, \underset{\sim}{u}\cdot\underset{\sim}{u} + \varepsilon\Psi(\underset{\sim}{u},s) - \underset{\sim}{w}_n(s)\cdot\underset{\sim}{u}$ has a unique point $\underset{\sim}{u}_\varepsilon(s)$ at which its gradient vanishes. The convexity of Ψ and (4.14) imply that

$$\Psi(\underset{\sim}{w}_n(s),s) - \Psi(\underset{\sim}{u}_\varepsilon(s),s) \geqslant \frac{\partial\Psi}{\partial\underset{\sim}{u}}(\underset{\sim}{u}_\varepsilon,s)\cdot(\underset{\sim}{w}_n(s) - \underset{\sim}{u}_\varepsilon(s)) \geqslant 0 \tag{4.15}$$

so that $\underset{\sim}{u}_\varepsilon \in F(n)$. Let Ω be an arbitrary measurable subset of $[0,1]$. Set

$$\underset{\sim}{u}(s) = \begin{cases} \underset{\sim}{u}_\varepsilon(s) & \text{for } s \in \Omega, \\ \underset{\sim}{w}_n(s) & \text{for } s \notin \Omega. \end{cases} \tag{4.16}$$

Then (4.15) implies that $\underset{\sim}{u} \in F(n)$. With this $\underset{\sim}{u}$, (4.13) reduces to

$$\int_{\Omega}\{\hat{\underset{\sim}{\sigma}}(\underset{\sim}{w}_n(s),s)+\underset{\sim}{\rho}[\underset{\sim}{w}_n](s)\}\cdot[\underset{\sim}{u}_{\varepsilon}(s)-\underset{\sim}{w}_n(s)]ds \geqslant 0 \quad \forall \text{ measurable } \Omega, \tag{4.17}$$

whence it follows that

$$\{\hat{\underset{\sim}{\sigma}}(\underset{\sim}{w}_n(s),s)+\underset{\sim}{\rho}[\underset{\sim}{w}_n](s)\}\cdot[\underset{\sim}{u}_{\varepsilon}(s)-\underset{\sim}{w}_n(s)] \geqslant 0 \quad \text{a.e.} \tag{4.18}$$

Equation (4.14) then implies that

$$\hat{\sigma}(\underset{\sim}{w}_n(s),s)\cdot\frac{\partial\Psi(\underset{\sim}{u}_{\varepsilon}(s),s)/\partial\underset{\sim}{u}}{|\partial\Psi(\underset{\sim}{u}_{\varepsilon}(s),s)/\partial\underset{\sim}{u}|} \leqslant |\rho[\underset{\sim}{w}_n](s)| \leqslant \text{const}\,[1+|\underset{\sim}{w}_n(s)|]. \tag{4.19}$$

Since (4.14) implies that $\underset{\sim}{u}_{\varepsilon}(s) \longrightarrow \underset{\sim}{w}_n(s)$ a.e. as $\varepsilon \longrightarrow 0$, we can let $\varepsilon \longrightarrow 0$ in (4.19) to get

$$\hat{\sigma}(\underset{\sim}{w}_n(s),s)\cdot\frac{\partial\Psi(\underset{\sim}{w}_n(s),s)/\partial\underset{\sim}{u}}{|\partial\Psi(\underset{\sim}{w}_n(s),s)/\partial\underset{\sim}{u}|} \leqslant \text{const}\,[1+|\underset{\sim}{w}_n(s)|]\,. \tag{4.20}$$

In view of (4.20), conditions (4.4) and (4.6) imply that there is an integer N such that

$$\Psi(\underset{\sim}{w}_n(s),s) \leqslant N \quad \text{a.e.} \tag{4.21}$$

Thus $\underset{\sim}{w}_n \in F(N)$ for all n. Now by letting Ω be an arbitrary measurable set and by taking $\underset{\sim}{u}(s) = \underset{\sim}{w}_n(s)$ on Ω^c, we find that (4.13) implies that

$$\{\hat{\underset{\sim}{\sigma}}(\underset{\sim}{w}_n(s),s)+\underset{\sim}{\rho}[\underset{\sim}{w}_n](s)\}\cdot[\underset{\sim}{u}(s)-\underset{\sim}{w}_n(s)] \geqslant 0 \quad \text{a.e.} \quad \forall\, u \in F(n). \tag{4.22}$$

Let $n > N$ and let $\underset{\sim}{a}$ be an arbitrary element of $\mathbb{R}^3$. Then (4.21) implies that $\underset{\sim}{u} = \underset{\sim}{w}_n + t\underset{\sim}{a} \in F(n)$ for sufficiently small t. We substitute this $\underset{\sim}{u}$ into (4.22) to obtain

$$\{\hat{\underset{\sim}{\sigma}}(\underset{\sim}{w}_n(s),s)+\underset{\sim}{\rho}[\underset{\sim}{w}_n](s)\}\cdot\underset{\sim}{a} \geqslant 0 \quad \text{a.e.} \quad \forall\, \underset{\sim}{a} \in \mathbb{R}^3, \tag{4.23}$$

whence follows (4.11). □

The rest of the regularity theory follows as in Section 3. (An alternative approach is given by Antman [6].) We note that (4.6) is a natural hypothesis for the use of the Poincaré-Bohl Theorem in finite dimensional degree theory. It can be used directly in the proof of the theorem of Brezis used in the proof. We also note that (4.6) affords a very simple and physically natural statement of coercivity.

5 CONCLUSION

On philosophical grounds one may object to the use of global existence theorems for problems from physics because they depend upon growth conditions restricting the behavior of large solutions: These conditions are completely inaccessible to experimental verification. This difficulty is somewhat mitigated by the mildness of the conditions we have imposed on the constitutive functions for large $|\underset{\sim}{u}|$ and for small $\delta(\underset{\sim}{u})$. One way to circumvent these problems is to use continuation methods that have been recently developed (cf. Rabinowitz [18], Alexander & Yorke [2]). In these methods the solution is shown to exist in the neighborhood of a known or trivial solution by some generalized implicit function theorem. By endowing such a solution with topological invariants, one may be able to show that the local solution can be extended to a continuum of solutions. Such methods are proving very useful for the determination of qualitative properties of solution branches for bifurcation problems of nonlinear elasticity (cf. Antman [9]).

In Sections 3 and 4 we used two different admissible classes, A and B, with B properly contained in A. It is conceivable that in a more complicated problem one could find a solution to a variational problem in B by the method of Section 4, but that this extremizer would fail to be a local minimizer on A because elements of $A\backslash B$ have been removed from competition. This fact would obstruct the study of the stability of motion of such minimizers (cf. Knops & Wilkes [15] and Browne [12]).

REFERENCES

1 N. I. Akhiezer, Lectures on the Calculus of Variations (in Russian). (1955) G.I.T.T.L. English transl. (1962), Blaisdell.

2 J. Alexander and J. Yorke, Global Bifurcation of Periodic Orbits, Am. J. Math., to appear.

3 S. S. Antman, Existence of Solutions of the Equilibrium Equations for Nonlinearly Elastic Rings and Arches, Indiana U. Math. J., 20, (1970), 281-302.

4 S. S. Antman, Existence and Nonuniqueness of Axisymmetric Equilibrium States of Nonlinearly Elastic Shells, Arch. Rational Mech. Anal., 40, (1971), 329-371.

5 S. S. Antman, The Theory of Rods, Handbuch der Physik, Vol. VIa/2, (1972), Springer, Verlag, 641-703.

6 S. S. Antman, Monotonicity and Invertibility Conditions in One-Dimensional Nonlinear Elasticity, Symp. on Nonlinear Elasticity (ed. R. W. Dickey), (1973), Academic Press 57-92.

7 S. S. Antman, Ordinary Differential Equations of One-Dimensional Nonlinear Elasticity I: Foundations of the Theories of Nonlinearly Elastic Rods and Shells, Arch. Rational Mech. Anal. 61, (1976), 307-351.

8 S. S. Antman, Ordinary Differential Equations of One-Dimensional Nonlinear Elasticity II: Existence and Regularity Theory for Conservative Problems, Arch. Rational Mech. Anal., 61, (1976), 353-393.

9 S. S. Antman, Bifurcation Problems for Nonlinearly Elastic Structures, Applications of Bifurcation Theory (ed. P. H. Rabinowitz), (1977), Academic Press, 73-125.

10 J. M. Ball, Convexity Conditions and Existence Theorems in Nonlinear Elasticity, Arch. Rational Mech. Anal., 63, (1977), 337-403.

11 H. Brezis, Équations et inéquations non linéaires dans les espaces vectoriels en dualité, Ann. Inst. Fourier, 18, (1968), 115-175.

12 R. Browne, Dynamic Stability of One-Dimensional Nonlinearly Visco-elastic Bodies, Dissertation, Univ. Maryland, (1976).

13 E. Hewitt and K. Stromberg, Real and Abstract Analysis, Springer, (1965).

14 T. J. R. Hughes, T. Kato and J. E. Marsden, Well-posed Quasi-linear Hyperbolic Systems, Arch. Rational Mech. Anal. 63(1977), 273-294.

15 R. J. Knops and E. W. Wilkes, Elastic Stability, Handbuch der Physik, Vol. VIa/3, (1973), Springer.

16 M. A. Krasnosel'skii and Ya. B. Rutitskii, Convex Functions and Orlicz Spaces (in Russian), (1958), Fizmatgiz; English transl. (1961), Noordhoff.

17 J.-L. Lions, Quelques méthodes de résolution des problèmes aux limites non linéaires, (1969), Dunod.

18 P. H. Rabinowitz, Some Aspects of Nonlinear Eigenvalue Problems, Rocky Mountain J. Math., 3, (1973), 161-202.

19 H. L. Royden, Real Analysis, 2nd ed., (1968), Macmillan.

20 C. Truesdell, Das ungelöste Hauptproblem der endlichen Elastizitätstheorie, Z.A.M.M., 36, (1956), 97-103.

21 M. M. Vainberg, Variational Methods for the Study of Nonlinear Operators (in Russian), English transl. (1964), Holden-Day, San Francisco.

22 C.-C. Wang and C. Truesdell, Rational Elasticity, (1973), Noordhoff.

Professor S. S. Antman, Department of Mathematics, University of Maryland, College Park, Maryland 20742, U.S.A.

Professor H. Brezis, Department of Mathematics, Université P. et M. Curie, 4 Place Jussieu, 75230 Paris Cedex 05, France.

J E MARSDEN AND T J R HUGHES

Topics in the mathematical foundations of elasticity

INTRODUCTION

We present some selected topics in elasticity theory from a point of view which aims at a healthy balance between geometry and analysis. Much of the work revolves around the Hamiltonian structure of elastodynamics. However, to keep the notes within reasonable bounds, we have not discussed to any extent our work on bifurcation theory (see Holmes and Marsden [128], Marsden [183] and Chillingworth and Marsden [42].)

The background assumed has been determined mostly by pedagogical factors. It was out of the question to develop the necessary machinery in geometry and differential forms but this can be found in Bishop and Goldberg [26] and Abraham and Marsden [1]. On the other hand, in order to make our presentation of existence and uniqueness theorems as complete as possible, the necessary facts about semigroups are proved, but those about elliptic operators are not. The latter is again for obvious pedagogical reasons.

The first three sections present standard topics found in continuum mechanics texts, but using the notation and concepts from modern differential geometry and nonlinear analysis (cf. Hughes and Marsden [132]). Already at this stage some of the results are new.

It is our conviction that not enough critical attention has been given to geometric concepts in continuum mechanics. Some confusion, even amongst some professionals, over ideas of covariance, objectivity, linearization, stress rates, conservation laws etc., has accompanied this neglect. However, these opening sections have to be taken for what they are: a more concise and elegant language for a well established discipline. It does not provide easy answers to hard analytical questions, but our personal experience has been that it helps to formulate things more efficiently and to get faster to the heart of the analysis. Others will find, quite understandably, that the investment required to learn the requisite calculus on manifolds simply isn't worth it.

Here is an example from Section 10 of how geometry can help guide the analysis. In 1966, Arnold [9] showed how the solutions of the Euler equations for a perfect fluid can be regarded as geodesics on the group of volume preserving diffeomorphisms. This is, in effect, a fancy way of setting up the equations in Lagrangian (material) coordinates. The remarkable thing, discovered by Ebin and Marsden [77], is that in this Lagrangian framework, the local existence theory is trivial since the differential equations become Lipschitz. This idea led to a number of important results such as the proof that the infinite Reynolds number limit ($\nu \longrightarrow 0$) is non-singular if boundaries are not present. Another example in the same context is Ebin's proof [76] of the convergence of compressible flow to incompressible flow as the compressibility tends to 0. Some, but not all of these results were subsequently proved using 'pure analysis'. However, in retrospect, the original geometric approach seems richer and more natural, especially to someone trained in material coordinates (as most elasticians are).

In elasticity, the analytical results obtained so far using geometric methods, are modest. However one important application is to the extention from the compressible to the incompressible case of the local dynamic existence theory. (See Section 10.) It seems likely that Ebin's results on the incompressible limit will also carry over. Ultimately, global methods in the calculus of variations due to Palais, Smale and Tromba (see Tromba [242]) should be useful for global results in elastostatics. (See Sections 8 and 11.)

The final section surveys a few known facts about existence and uniqueness theory (Ball [12] and Hughes, Kato and Marsden [129]), and discusses some related problems in stability theory. A brief sketch is given of an approach to the existence and nonuniqueness theorems of Stoppelli [236] by means of generic bifurcation theory in the sense of Chow, Hale and Mallet-Paret [47].

Many of the results presented here are new, but we have included many that are standard as well in order to make the notes as accessible as possible. Further, for the sake of timeliness, a number of ideas which are not yet fully developed or polished are also included. Needless to say, our work leans heavily on that of the masters, in particular on that of Truesdell and Noll [248] and Rivlin [220,221].

Our thanks are extended to Robin Knops and John Ball for their hospitality at Heriot-Watt University. Conversations with them were an invaluable assistance. Helpful comments by S. Antman, J. Carr, C. Dafermos, M. Gurtin, R. N. Hills, R. Muncaster and N. Wilkes are gratefully acknowledged. Some of the ideas presented in Section 3 on covariant constitutive theory are still in a preliminary state and were obtained in collaboration with M. Gurtin. We also thank C. Navarro for contributing the appendix to Section 7 on semigroup techniques applied to linear elastodynamics. Our presentation of semigroup theory has been influenced by course notes of T. Kato and P. Chernoff.

1 KINEMATICS

By a body we mean a smooth manifold M, possibly with boundary. Points in M are denoted by $X \in M$ and local coordinates by $\{X^A\}$. The tangent space to M at X is denoted $T_X M$.

We shall not worry about differentiability in this section. Manifolds and maps will be assumed to be as differentiable as needed for the results to make sense. Of course for the later existence theory this will be nailed down precisely.

The space in which we imagine the body M to move is denoted by N, again a smooth manifold. Points in N are denoted by $x \in N$ and local coordinates in N by $\{x^a\}$.

By a configuration or a deformation of M we mean a map

$$\phi : M \longrightarrow N.$$

We shall always assume ϕ is regular i.e., is a diffeomorphism onto its image. One also says ϕ is an embedding of M into N. The set of all configurations is denoted $\mathscr{C}$, and is called the configuration space. It can be shown to be a smooth infinite dimensional manifold using the methods in, for example, Ebin and Marsden [77]. We won't really need this fact until later (See especially Sections 4 and 10).

A motion of M is a curve $\phi(t)$ in $\mathscr{C}$. We write $\phi(t,X) = \phi_t(X)$ for this curve evaluated at $X \in M$, and let $x = \phi(t,X)$.

The material velocity V of a motion is defined by

$$V(t,X) = \frac{\partial}{\partial t}\phi(t,X).$$

We may regard V as a tangent vector to the curve $\phi(t)$ in $\mathscr{C}$. From its definition, V_t is a vector field over ϕ_t, i.e., $V_t : M \longrightarrow TN$,

$V_t(X) \in T_{\phi(X)}N$ for each $X \in M$. (See Fig. 1.1) The spatial velocity v_t is the vector field in N defined on $\phi_t(M)$ by

$$v_t = V_t \circ \phi_t^{-1}$$

We call ϕ_0 the reference configuration. Usually we can assume that ϕ_0 = identity.

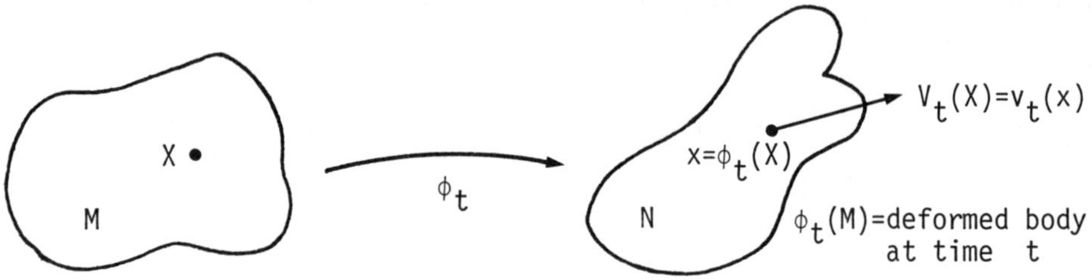

Fig. 1.

We write $F = T\phi$ or $D\phi$, the tangent i.e., derivative of the configuration ϕ. Thus for each $X \in M$, $x = \phi(X)$, $F(X): T_X M \longrightarrow T_x N$ is a linear map. We call F the deformation gradient. It is an example of a two point tensor; i.e., multilinear maps from copies of $T_X M$, $T_X^* M$, $T_x N$ and $T_x^* N$ to the reals $\mathbb{R}$.

In coordinates, F has components

$$F^a{}_A = \frac{\partial \phi^a}{\partial X^A}.$$

We assume there is a Riemannian metric g on N and a Riemannian metric G on M. Their respective Christoffel symbols in coordinates are denoted γ^a_{bc} and Γ^A_{BC}.

The covariant derivative of the material velocity V of a motion with respect to t is called the material acceleration and is denoted by A. In coordinates,

$$A^a = \frac{\partial V^a}{\partial t} + \gamma^a_{bc} V^b V^c ,$$

where all quantities are evaluated at the appropriate points (i.e., A^a, V^a at (t,X) and γ^a_{bc} at $x = \phi_t(X)$). The spatial acceleration is defined by

$$a_t = A_t \circ \phi_t^{-1} = \frac{\partial v}{\partial t} + \nabla_v v.$$

In coordinates,

$$a^a = \frac{\partial v^a}{\partial t} + \frac{\partial v^a}{\partial x^b} v^b + \gamma^a_{bc} v^b v^c .$$

The extra term $\frac{\partial v^a}{\partial x^b} v^b$ arises by using the chain rule to compare $\partial V^a/\partial t$ and $\partial v^a/\partial t$.

The material deformation tensor (or right Cauchy-Green tensor) C is defined by

$$C = \phi^*(g)$$

where ϕ^* denotes pull back of tensors by ϕ. In coordinates,

$$C_{AB} = g_{ab} F^a{}_A F^b{}_B .$$

Clearly C_{AB} is symmetric and positive definite (since g_{ab} is and $F^a{}_A$ is invertible).

The spatial deformation tensor (or left Cauchy-Green tensor) is similarly defined by

$$c = \phi_*(G),$$

the push-forward of G by ϕ. In coordinates,

$$c_{ab} = G_{AB} (F^{-1})^A{}_a (F^{-1})^B{}_b .$$

We write $G^\sharp$ for the tensor inverse to G and $g^\sharp$ for that of g. In coordinates $G^\sharp$ has components G^{AB}, the inverse matrix of G_{AB}. We use these to raise and lower indices on tensors, thereby forming associated tensors in the usual way. It is important to distinguish tensors from their associated tensors since pull back and push forward do not commute with this operation.

In fact we define the Finger deformation tensors respectively to be

$$B = \phi^*(g^\sharp), \quad b = \phi_*(G^\sharp).$$

Let K denote the curvature tensor of $\phi^*(g)$ and L that of g. From covariance principles in geometry we know that

$$K = \phi^* L.$$

In particular if g is flat, $K = 0$. These are usually called the compatability conditions. Conversely if, in Euclidean space we are given an F and form K and find $K = 0$, then F arises from a ϕ. This follows from the result that a metric with zero curvature has constant components in normal coordinates.

It is common to decompose F according to the polar decomposition

$$F = R \circ U = V \circ R$$

where U, V are positive definite and symmetric (the square roots of C, c) and R is orthogonal. We remark that this makes perfectly good sense on manifolds: notice that

$$R : TM \longrightarrow TN,$$
$$U : TM \longrightarrow TM,$$
$$V : TN \longrightarrow TN.$$

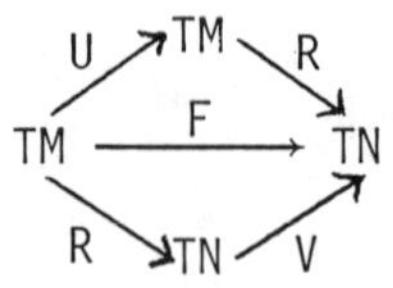

One calls R the <u>rotation tensor</u> and U,V the right and left <u>stretch tensors</u>.

The <u>Lie derivative</u> $L_v h$ of a possibly time dependent tensor field h on N with respect to the spatial velocity is defined by

$$\phi_t^*(L_v h) = \frac{\partial}{\partial t}(\phi_t^* h).$$

We write $L_v = \frac{\partial}{\partial t} + \mathcal{L}_v$ so that L_v and $\mathcal{L}_v$ coincide on time independent tensor fields. This is the normal way geometers measure the rate of change of a tensor with respect to a vector field. Note that it doesn't involve the metric, although its componential expression may be written in terms of either ordinary or covariant derivatives.

If the indices on h are raised or lowered one obtains different expressions, since Lie differentiation does not commute with this operation. If h is taken to be the stress tensor or the stress tensor density, these possibilities result in the various expressions for stress rates proposed by Oldroyd, Jaumann and Truesdell. We shall study these expressions and "objective rates" in general as an application of Lie derivatives as soon as a little more notation is reviewed.

The <u>rate of deformation tensor</u> in spatial coordinates is defined by

$$d = \tfrac{1}{2}\mathcal{L}_v g,$$

or

$$d_{ab} = \tfrac{1}{2}(v_{a|b} + v_{b|a})$$

and in material coordinates by,

$$D = \tfrac{1}{2}\frac{\partial}{\partial t}C.$$

Thus,

$$\phi^*d = D.$$

We assume M and N are oriented and write dv for the volume element on N. Thus, in coordinates

$$dv = \sqrt{\det g_{ab}}\, dx^1 \wedge \ldots \wedge dx^n, \quad n = \dim N.$$

Similarly,

$$dV = \sqrt{\det G_{AB}}\, dX^1 \wedge \ldots \wedge dX^m, \quad m = \dim M,$$

is the volume element on M. The <u>Jacobian</u> J of ϕ is defined by

$$\phi^*dv = JdV;$$

explicitly,

$$J(X) = \frac{\partial(\phi^1, \ldots, \phi^n)}{\partial(X^1, \ldots, X^n)} \frac{\sqrt{\det g_{ab}}}{\sqrt{\det G_{AB}}} .$$

This is defined only if $n = m$. In this case, let a positive function $\rho(t,x)$, $x \in \phi_t(M)$ be given. We say it satisfies <u>conservation of mass</u> if

$$\frac{d}{dt} \int_{\phi_t(U)} \rho(t,x)dv = 0$$

for every smooth open submanifold $U \subset M$.

Define the <u>divergence</u> div v of v, by

$$\mathcal{L}_v dv = (\mathrm{div}\, v)dv;$$

i.e.,

$$\frac{\partial J}{\partial t} = (\mathrm{div}\, v)J$$

In coordinates we get the usual formula:

$$\operatorname{div} v = v^a{}_{|a} = \frac{1}{\sqrt{\det g_{cd}}} \frac{\partial}{\partial x^a} (\sqrt{\det g_{cd}}\, v^a).$$

Then conservation of mass is equivalent to the continuity equation

$$\frac{\partial \rho}{\partial t} + d\rho \cdot v + \rho \operatorname{div} v = 0.$$

Indeed, this follows by changing variables:

$$\frac{d}{dt} \int_{\phi_t(U)} \rho(t,x) dv = \frac{d}{dt} \int_U \rho(t,\phi_t(X)) J\, dV = \int_U (\frac{\partial \rho}{\partial t} + \rho \operatorname{div} v) J\, dV.$$

Since U is arbitrary, and the integrand is tacitly assumed to be continuous, conservation of mass is equivalent to the continuity equation.

We write $\frac{\partial \rho}{\partial t} + d\rho \cdot v = \dot{\rho}$ and call it the <u>material derivative</u>. Note that

$\dot{\rho}_t = (\frac{\partial}{\partial t} \rho_t \circ \phi_t) \circ \phi_t^{-1}$, which is used to define the material derivative of a general spatial tensor field.

Conservation of mass is more interesting if $\dim M \neq \dim N$. Take the case of shells as an illustration. Here, $\dim M = \dim N - 1$. Let k denote the second fundamental form (extrinsic curvature) of $\phi_t(M)$ in N. (For a surface in space, $\operatorname{tr} k$ is the mean curvature and $\det k$ is the Gaussian curvature.) Now write $v = v_{\parallel} + v_n n$, where $v_{\parallel} \in T(\phi_t(M))$, i.e., is tangent to the shell, and n is a unit normal, relative to a given orientation.

The equation of continuity then reads

$$\dot{\rho} + \rho(\operatorname{div} v_{\parallel}) + \rho v_n \operatorname{tr} k = 0 .$$

We leave the proof as an exercise for the geometrically inclined reader.

As a first, and rather simple application of geometric ideas, specifically the Lie derivative, we shall consider <u>objective rates</u>.

Let $\underset{\sim}{t}$ be a given symmetric covariant (indices down) two tensor, the stress tensor, say. Let $\underset{\sim}{t}_1, \underset{\sim}{t}_2, \underset{\sim}{t}_3$ be the three associated tensors with indices raised by the metric $\underset{\sim}{g}$ and let $\underset{\sim}{t}_4 = \underset{\sim}{t}$. In coordinates $\{x^a\}$ on N,

$$\underset{\sim}{t}_1 = (t^{ab}), \quad \underset{\sim}{t}_2 = (t_a{}^b), \quad \underset{\sim}{t}_3 = (t^a{}_b), \quad \underset{\sim}{t}_4 = (t_{ab}).$$

On noting that the Lie derivative does not commute with raising or lowering indices, we get four different formulas:

$$(L_{\underset{\sim}{v}} \underset{\sim}{t}_1)^{ab} = \dot{t}^{ab} - t^{cb} v^a{}_{|c} - t^{ac} v^b{}_{|c},$$

$$g^{ac} (L_{\underset{\sim}{v}} \underset{\sim}{t}_2)_c{}^b = \dot{t}^{ab} - t^{ad} v^b{}_{|d} + t^{ab} v_{d|}{}^a,$$

$$(L_{\underset{\sim}{v}} \underset{\sim}{t}_3)^a{}_c \, g^{cb} = \dot{t}^{ab} - t^{ab} v^a{}_{|d} + t^{ad} v_{d|}{}^b,$$

and

$$g^{ac} (L_v t\ )_{cd} \, g^{db} = t^{ab} + t^{cb} v_c{}^{|a} + t^{ac} v_c{}^{|b},$$

while for the density $\underset{\sim}{t}_1 \otimes dv$ where dv is the volume element for the metric $\underset{\sim}{g}$, we get

$$(L_v(\underset{\sim}{t}_1 \otimes dv))^{ab} = ((L_{\underset{\sim}{v}} \underset{\sim}{t}_1)^{ab} + t^{ab} \operatorname{div} v) dv.$$

We shall show shortly that it is not an accident that the so-called "objective fluxes" (the right-hand sides above) turn out to be Lie derivatives with respect to the velocity.

The tensor $L_{\underset{\sim}{v}} \underset{\sim}{t}_1$ has been associated with the name Oldroyd and $L_v(\underset{\sim}{t}_1 \otimes dv)$ with the name Truesdell. We see that all of these tensors are different manifestations of the Lie derivative of $\underset{\sim}{t}$.

Any linear combination of the preceding formulas will also qualify as an "objective flux", e.g.,

$$\tfrac{1}{2}((L_{\underset{\sim}{v}}t_3)^a{}_c g^{cb} + g^{ac}(L_{\underset{\sim}{v}}t_2)_c{}^b) = \dot{t}^{ab} + t^{ad}\omega_d{}^b - t^{db}\omega^a{}_d,$$

where $\omega^a{}_b$ are associated components of the spin $2\omega_{ab} = v_{a|b} - v_{b|a}$; this tensor is associated with the name Jaumann.

(We note in passing that, like the Lie derivative in general, the right-hand sides may be equally well expressed without using covariant derivatives.)

To complete the discussion we need to define the term "objective" in the language of differential geometry. These notions will also be needed in the next section.

Definition Let $\underset{\sim}{t}$ be a tensor field (or tensor density) on a manifold N and ξ a diffeomorphism of N to N. We say that the push-forward $\underset{\sim}{t}' = \xi_* \underset{\sim}{t}$ is the objective or spatial transformation of $\underset{\sim}{t}$, i.e., $\underset{\sim}{t}$ transforms in the usual way under the map ξ.

Proposition Let ϕ_t be a regular motion of M in N with velocity field v_t (spatial velocity). Let ξ_t be a motion of N in N' and let $\phi'_t = \psi_t \circ \phi_t$ be the superposed motion of M in N'.

Let $\underset{\sim}{t}$ be a given time-dependent tensor field on N and let

$$\underset{\sim}{t}' = \xi_* \underset{\sim}{t},$$

i.e., transform $\underset{\sim}{t}$ objectively.

Let v' be the velocity field of ϕ'_t. Then

$$L_{v'}\underset{\sim}{t}' = \xi_*(L_{\underset{\sim}{v}}t),$$

i.e., "objective tensors (or tensor densities) have objective Lie derivatives".

Proof. We first note that

$$v'_t = w_t + \xi_{t*}v_t$$

where w_t is the spatial velocity of ξ_t. This follows by differentiating $\phi'_t(X) = \xi_t(\phi_t(X))$ in t. (As can be seen, v is not objective.)

Let $\xi_{r,s} = \xi_r \circ \xi_s^{-1}$ be the time dependent flow of $\underset{\sim}{w}$. Then,

$$L_{v'} \underset{\sim}{t}' = L_{w+\xi_* v}(\xi_* \underset{\sim}{t})$$

$$= \mathcal{L}_{w+\xi_* v}(\xi_* \underset{\sim}{t}) + \frac{\partial}{\partial t}(\xi_* \underset{\sim}{t})$$

$$= \xi_*(\mathcal{L}_v \underset{\sim}{t}) + \mathcal{L}_w(\xi_* \underset{\sim}{t}) + \frac{\partial}{\partial t}(\xi_* \underset{\sim}{t})$$

$$= \xi_*(\mathcal{L}_v \underset{\sim}{t}) + L_w(\xi_* \underset{\sim}{t})$$

$$= \xi_*(\mathcal{L}_v \underset{\sim}{t}) + \frac{d}{dr}\, \xi^*_{r,t}(\xi_{r*} \underset{\sim}{t}_r)\Big|_{r=t}$$

$$= \xi_*(\mathcal{L}_v \underset{\sim}{t}) + \frac{d}{dr}(\xi_r \circ \xi_t^{-1})^*(\xi_{r*} \underset{\sim}{t}_r)\Big|_{r=t}$$

$$= \xi_*(\mathcal{L}_v \underset{\sim}{t}) + \frac{d}{dr}\, \xi_{t*} \underset{\sim}{t}_r\Big|_{r=t}$$

$$= \xi_*\left(\mathcal{L}_v \underset{\sim}{t} + \frac{d}{dr}\, \underset{\sim}{t}_r\Big|_{r=t}\right)$$

$$= \xi_*(L_v \underset{\sim}{t}). \qquad \square$$

This can also be verified by a direct, although somewhat messy, coordinate calculation.

As a second application of geometric ideas to kinematics, we consider another important operation which will be of use in subsequent sections. This is the Piola transform. Its main use is to connect the material and spatial pictures. In doing so, the Piola identity plays a central role.

Let y be a vector field on N, Y be a vector field on M and assume $\dim N = \dim M$. We say Y is the <u>Piola transform</u> of y if

$$Y = J\phi^* y.$$

We remark that this is equivalent to $\phi^*(i_y dv) = i_y dV$ where $i_y dv$ denotes interior product ($= y \lrcorner\, dv$). This is because,

$$\phi^*(i_y dv) = i_{\phi^* y}\phi^* dv = i_{\phi^* y} J dV = i_{J\phi^* y} dV.$$

We also remark that if da denotes the area element on a hypersurface and n is its oriented unit normal, then on this hypersurface

$$i_y dv = \langle y, n \rangle\, da.$$

The proof is a simple computation.

<u>Proposition</u> (The Piola identity) $\mathrm{DIV}\, Y = J(\mathrm{div}\, y \circ \phi)$

<u>Proof 1</u>. $\displaystyle\int_{\partial U} \langle Y, N \rangle\, dA = \int_U \mathrm{DIV}\, Y\, dV$ for any open subset $U \subset M$ with piecewise C^1 boundary ∂U, with a similar formula for y. But by change of variables

$$\int_{\partial U} \langle Y, N \rangle\, dA = \int_{\partial U} i_y dV = \int_{\partial\phi(U)} i_y dv = \int_{\partial\phi(U)} \langle y, n \rangle\, da.$$

Thus $\displaystyle\int_U \mathrm{DIV}\, Y\, dV = \int_{\phi(U)} \mathrm{div}\, y\, dv = \int_U J(\mathrm{div}\, y \circ \phi) dv$ so as U is arbitrary, the result follows. □

Another way of saying this identity is $DIV(JF^{-1}) = 0$, i.e., $(J(F^{-1})^A{}_a)_{|A} = 0$. This follows from what we proved: viz.,

$$(J(F^{-1})^A{}_a y^a)_{|A} = Jy^a{}_{|a}$$

holds for all y.

<u>Proof 2</u>. We compute directly, using differential forms:

$$\begin{aligned}(DIV\, Y)dV &= L_y dV \\ &= d(i_y dV)\end{aligned}$$

(By the general formula

$$L_y\alpha = i_y d\alpha + di_y\alpha$$

and the fact that d of an n-form is zero.)

Thus

$$\begin{aligned}(DIV\, Y)dV &= d\phi^*(i_y dv) \\ &= \phi^* d(i_y dv) \qquad \text{(pull-back and } d \text{ commute)} \\ &= \phi^*(div\, y\; dv) \\ &= J(div\, y \circ \phi)dV \qquad \text{(definition of } J)\end{aligned}$$

and so

$$DIV\, Y = J(div\, y \circ \phi). \quad \square$$

Our final application is to the well-known criterion of Vainberg [250] for a set of equations to be derivable from a variational principle, i.e., to be Euler-Lagrange equations. The same result appears, in different form in many places. For example one can ask, which is essentially the same thing, when a given vector field is Hamiltonian. The answer (see Chernoff and Marsden [39] p.78 and Sections 6 and 9 below) was known to Poincaré and Cartan. Its application to elasticity yields the well known result that the elasticity tensor is symmetric if and only if it derives from an internal energy function. We shall see this application later.

Let X, Y be Banach spaces with Y densely and continuously included in X. Let $\langle\, ,\rangle$ be a continuous bilinear form on X and let $A: Y \longrightarrow X$ be a given (nonlinear) operator. The Fréchet derivative of A at x is denoted $DA(x)$.

<u>Definition</u> We say A is a <u>potential operator</u> if there is a function $L: Y \longrightarrow \mathbb{R}$ such that

$$dL(x)\cdot v = \langle A(x), v\rangle$$

for all $x, v \in Y$.

The equation $A(x) = 0$ represents in abstract form the Euler-Lagrange equations for $x \in Y$.

<u>Proposition</u> A given operator A is a potential operator if and only if for each $x \in Y$, and v_1 and $v_2 \in Y$,

$$\langle DA(x)\cdot v_1, v_2\rangle = \langle DA(x)\cdot v_2, v_1\rangle .$$

If $\langle\, ,\rangle$ is symmetric, this equivalent to saying $DA(x)$ is a symmetric linear operator on X (with domain Y).

<u>Proof</u>. Consider the one form $\alpha(x)\cdot v = \langle A(x), v\rangle$ on Y. Then A is a potential operator if and only if α is exact. By the Poincaré lemma, this is the case if and only if $d\alpha = 0$. But by the coordinate formula for

exterior derivative (the "curl" in this case)†,

$$d\alpha(x)\cdot(v_1,v_2) = \langle DA(x)\cdot v_1, v_2\rangle - \langle DA(x)\cdot v_2, v_1\rangle$$

so the result follows immediately. □

The proof of the Poincaré lemma also gives us a formula for L in terms of A:

$$L(x) = \int_0^1 \langle A(\tau x), x\rangle d\tau.$$

Using this formula the proposition may be verified directly.

2 BALANCE LAWS

We shall begin by reviewing standard results on the foundations of continuum theory. The balance laws in integral form are often postulated as the basic axioms of the theory. Unfortunately, such a key law as balance of momentum is not a covariant statement. It is the authors' belief that if a theory pretends to be fundamental (as opposed to particular examples), it must have a covariant formulation. In other words, on any manifold in any coordinate system it must be clear how to formulate the theory. Thus, theories which are given explicitly in Euclidean space in Euclidean coordinates are not covariant in this sense. (Changing coordinates in Euclidean space does not tell you how to formulate the theory on a three-sphere or for shells, etc. Some exotic materials such as liquid crystals may be best formulated in spaces which are not Euclidean, so the reasons for demanding covariance are local as well as global.)

Demanding covariance gives, we believe, some fresh insight into balance

---

†See, Lang [167] or Abraham and Marsden [1], Table 10.1.

of energy and the relationship between the first and second laws of thermodynamics.

This point of view is not our personal prejudice. F. Dyson [74] has made the same point relative to quantum theory in crystal clear terms:

> "The most glaring incompatibility of concepts in contemporary physics is that between Einstein's principle of general coordinate invariance and all the modern schemes for a quantum-mechanical description of nature. Einstein based his theory of general relativity on the principle that God did not attach any preferred labels to the points of space-time. This principle requires that the laws of physics should be invariant under the Einstein group E, which consists of all one-to-one and twice-differentiable transformations of the coordinates. By making full use of the invariance under E, Einstein was able to deduce the precise form of his law of gravitation from general requirements of mathematical simplicity without any arbitrariness. He was also able to reformulate the whole of classical physics (electromagnetism and hydrodynamics) in E-invariant fashion, and so determine unambiguously the mutual interactions of matter, radiation and gravitation within the classical domain. There is no part of physics more coherent mathematically and more satisfying aesthetically than this classical theory of Einstein based upon E-invariance.
>
> On the other hand, all the currently viable formalisms for describing nature quantum-mechanically use a much smaller invariance group. The analysis of Bacry and Lévy-Leblond indicates the extreme range of quantum-mechanical kinematical groups that have been contemplated. In practice all serious quantum-mechanical theories are based either on the Poincaré group P or the Galilei group G. This means that a

class of preferred inertial coordinate-systems is postulated a priori, in flat contradiction to Einstein's principle. The contradiction is particularly uncomfortable, because Einstein's principle of general coordinate invariance has such an attractive quality of absoluteness. A physicist's intuition tells him that, if Einstein's principle is valid at all, it ought to be valid for the whole of physics, quantum-mechanical as well as classical. If the principle were not universally valid, it is difficult to understand why Einstein achieved such deeply coherent insights into nature by assuming it to be so."

As in Section 1, let M and N be manifolds representing the 'body' and 'space' respectively and let ϕ_t be a (regular) motion of M in N with spatial velocity v and material velocity V. The manifolds M and N carry metrics G and g and dV and dv denote the respective volume elements.

Lemma (Transport Theorem) Let $f(x,t)$ be a C^1 scalar function of $x \in \phi_t(M)$ and t. Let $U \subset M$ be a bounded open set with piecewise C^1 boundary (hereafter called a 'nice' region). Then

$$\frac{d}{dt}\int_{\phi_t(U)} f(x,t)\,dv = \int_{\phi_t(U)} (\dot{f} + f \operatorname{div} v)\,dv.$$

Proof. By change of variables and the formula $\frac{\partial J}{\partial t} = (\operatorname{div} v \circ \phi_t)J$ from Section 1, we have

$$\frac{d}{dt}\int_{\phi_t(U)} f\,dv = \frac{d}{dt}\int_U (f \circ \phi_t)J\,dv$$

$$= \int_U \left(\frac{\partial}{\partial t}(f \circ \phi_t)J + (f \circ \phi_t)J \operatorname{div} v\right)dv$$

$$= \int_{\phi_t(U)} ((\frac{\partial}{\partial t} f \circ \phi_t) \circ \phi_t^{-1} + f \operatorname{div} v) dv.$$

But $\dot{f} = \left[\frac{\partial}{\partial t}(f \circ \phi_t)\right] \circ \phi_t^{-1}$ by definition of the material derivative

$(\dot{f} = \frac{\partial f}{\partial t} + df \cdot v)$. □

This lemma as it stands makes perfectly good sense on any manifold.

If $a(x,t)$ and $b(x,t)$ are scalar functions and $c(t,x)$ a vector field, they are said to satisfy the <u>master balance law</u> if

$$\frac{d}{dt} \int_{\phi_t(U)} a(x,t) dv = \int_{\phi_t(U)} b(x,t) dv + \int_{\partial\phi_t(U)} \langle c(x,t), n \rangle da$$

for all nice regions $U \subset M$, where n is the unit outward normal on $\partial\phi_t(U) = \phi_t(\partial U)$ and da is the corresponding area element. If $=$ is replaced by $\geqslant$, we refer to this as the <u>master balance inequality</u>.

<u>Proposition</u> If a, b, c are C^1 then the master balance law (resp. inequality) is equivalent to the identity

$$\dot{a} + a \operatorname{div} v = b + \operatorname{div} c$$

(resp. $\geqslant$). If ρ satisfies conservation of mass, then ρa, ρb, c satisfy the master balance law (resp. inequality) if and only if

$$\rho\dot{a} = \rho b + \operatorname{div} c.$$

<u>Proof</u>. This is a consequence of the transport theorem, the arbitrariness of U and the continuity equation $\dot{\rho} + \rho \operatorname{div} v = 0$. □

This can be done equally well in material coordinates. Namely, if we let $\rho_0 A$, $\rho_0 B$ and C satisfy

$$\frac{d}{dt}\int_U \rho_0 A\, dV = \int_U \rho_0 B\, dV + \int_U \langle C,N\rangle\, dA,$$

the localization becomes

$$\rho_0 \frac{\partial A}{\partial t} = \rho_0 B + \mathrm{DIV}\, C.$$

If $a_t = A_t \circ \phi_t^{-1}$, $b_t = B_t \circ \phi_t^{-1}$ and c is the Piola transform of C then this equation is equivalent to $\rho \dot{a} = \rho b + \mathrm{div}\, c$. This is seen by using the equation $\rho_0 = J(\rho \circ \phi_t)$ and the Piola identity.

<u>Cauchy's theorem</u> states that if c is a scalar function of x,t and n and if a,b,c satisfy the master balance law (or inequality), then $c(x,t,n)$ $c(x,t,n) = \langle c(x,t),n\rangle$ for a vector field $c(x,t)$. We refer to standard texts for the proof, noting that it is valid on general manifolds. (For the technically sharpest version available, see Gurtin and Martins [112]).

Given a motion of a body, we assume there exists a vector function (the Cauchy traction vector) τ depending on t,x and a unit vector n at x such that τ measures the force (per unit area orthogonal to n) of contact between two parts of the body.

We assume then that <u>balance of momentum</u> holds: for all (nice) $U \subset M$,

$$\frac{d}{dt}\int_{\phi_t(U)} \rho v\, dv = \int_{\phi_t(U)} \rho b\, dv + \int_{\partial\phi_t(U)} \tau\, da$$

where b is a (given) external force field. We shall criticize this assumption shortly.

By Cauchy's theorem, we can write

$$\tau(t,x,n) = \langle \underset{\sim}{t}(t,x),n\rangle$$

for a 2-contravariant (indices up) tensor field $\underset{\sim}{t}$, the <u>Cauchy stress tensor</u>. In coordinates,

$$\tau^a = n_b t^{ab}.$$

From the localization of the master balance law we find that balance of momentum is equivalent to the basic equations of motion

$$\rho a = \rho b + \operatorname{div} \underset{\sim}{t}.$$

Of course for these to be formally well-posed, we have to say how $\underset{\sim}{t}$ depends on the motion. This is the subject of constitutive theory (see Section 3).

We let T be the Piola transform on the first index of $\underset{\sim}{t}$; i.e.,

$$T^{Aa} = J(F^{-1})^A{}_b t^{ba}.$$

T is called the first Piola-Kirchhoff stress tensor. Then the material form of the equations of motion becomes

$$\rho_0 A = \rho_0 B + \mathrm{DIV}\, T$$

where $B = b \circ \phi_t^{-1}$.

Another important tensor is the second Piola-Kirchhoff stress tensor defined as the pull-back by ϕ_t of the second index of† T; i.e.,

$$P^{AB} = T^{Aa}(F^{-1})^B{}_a.$$

In $\mathbb{R}^3$ one also postulates balance of moment of momentum, i.e., for all (nice) $U \subset M$,

$$\frac{d}{dt}\int_{\phi_t(U)} \rho(x \times v)\, dv(x) = \int_{\phi_t(U)} \rho(x \times b)\, dv(x) + \int_{\partial\phi_t(U)} x \times \tau\, da(x).$$

---

†If one pulls back the covariant (indices down) form of t by ϕ, one gets the convected stress tensor, while if one pulls back by the rotational part R of F, one gets the co-rotational stress tensor, etc.

Assuming balance of momentum, it is an easy matter to check that this is equivalent to symmetry of $\underset{\sim}{t}$: $t^{ab} = t^{ba}$. Equivalently, P is symmetric: $P^{AB} = P^{BA}$. (Notice that it does not make tensorial sense to ask that T be symmetric.)

These two postulates are usually taken as the basic principles underlying continuum mechanics. We shall now criticize this point of view.

The main objection is that balance of momentum does not make covariant sense. Indeed it must be postulated in a Euclidean (or inertial) frame. In any other frame it will not look the same. On a general manifold it does not make sense.† Therefore, if we subscribe to Dyson's point of view, we must reject it as a fundamental postulate. (Of course the study of specific models is another matter.)

Balance of moment of momentum, of course, depends explicitly on $\mathbb{R}^3$ so is à priori objectionable on covariant grounds. Rod and shell theories also use these postulates, but they must be modified to take into account the particular geometry. This is another manifestation of the non-covariance of the fundamental laws.

Since most engineering mechanics occurs in $\mathbb{R}^3$, these objections are not relevant for practical considerations. However it is naive to think that $\mathbb{R}^3$ will cover all possibilities. For example if one wants to treat Cosserat (directed) continua without using higher order theory, one can do so with the theory at hand, but we must replace the containing space $\mathbb{R}^3$ by something more complicated. For example, for inextensible undirected rods, one uses $N = \mathbb{R}^3 \times \mathbb{P}^2$,

---

†One could attempt to integrate vector fields on a Riemannian manifold using parallel translation, but this only leads to conditions dependent on which point the vectors are translated to.

where $\mathbb{P}^2$ is real projective two space.[†] Again, in general relativity, these covariance questions are crucial for elasticity since inertial frames may not be at our disposal (see, eg., Carter and Quintana [36] and Maugin [188]).

There are two ways (at least) of making the foundations of continuum mechanics covariant. One way is to assume a Hamiltonian or Lagrangian structure and treat it as a classical field theory. This will be done in Sections 9 and 10. Another is to base it on a balance of energy principle and a covariant version of the Green-Rivlin [101] invariance assumption.

We shall first review the Green-Rivlin arguments.

Let $\rho(t,x)$, $b(t,x)$, $h(t,x,n)$, $\tau(t,x,n)$, $e(t,x)$ and $r(t,p)$ be given functions on $N = \mathbb{R}^3$, h and τ depending on a unit vector n. Here, h is the scalar heat flux, e the internal energy and r the heat source.

These functions are said to satisfy the <u>balance of energy</u> principle if, for all (nice) $U \subset M$,

$$\frac{d}{dt}\int_{\phi_t(U)} \rho(e + \tfrac{1}{2}\langle v,v\rangle)dv = \int_{\phi_t(U)} \rho(\langle b,v\rangle + r)dv + \int_{\partial\phi_t(U)} (\langle \tau,v\rangle + h)da,$$

where τ and h are evaluated on the unit outward normal n of $\partial\phi_t(U)$.

Although it can be derived,[*] let us assume at the outset, for simplicity, that τ and h have the form $\tau = \langle \underset{\sim}{t},n\rangle$ and $h = -\langle q,n\rangle$.

From the master balance law in localized form, a straightforward computation shows that this is equivalent to the local form

$$\rho(\dot{e} + \langle v,a\rangle) + (\dot{\rho} + \rho\,\mathrm{div}\,v)(e + \tfrac{1}{2}\langle v,v\rangle) = \rho r + \rho\langle b,v\rangle - \mathrm{div}\,q + \mathrm{div}(\underset{\sim}{t}\cdot v).$$

[†]Details of this sort of example are found in Naghdi [194]. See also Hughes and Marsden [132].

[*]One first obtains balance of momentum using translation invariance (see below) and hence $\tau = \langle t,n\rangle$; then by Cauchy's theorem applied to balance of energy, $h = \langle q,n\rangle$ follows.

No other balance principles (including conservation of mass) are used here. If the other balance principles are used, it simplifies to the <u>local energy balance</u>

$$\rho\dot{e} + \operatorname{div} q = \underset{\sim}{t}:d + \rho r$$

where $\underset{\sim}{t} : d = t^{ab} d_{ab}$.

Now let $\psi(t,x) = c(t) + Q(t)x$ denote a time-independent rigid motion of $\mathbb{R}^3$; i.e., $Q(t)$ is a 3×3 orthogonal matrix of determinant $+1$, and $\underset{\sim}{c}(t)$ is a vector.

Consider the new motion of M defined by

$$\phi'(t,X) = \psi(t,\phi(t,X)).$$

This has velocity

$$v'(t,x') = w(t,x) + Q(t)v(t,x)$$

where $x' = \phi'(t,X)$ and

$$w(t,x) = \frac{dc}{dt} + \frac{dQ}{dt}x.$$

Let $a'(t,x')$ be the acceleration of the new motion. With this new motion, associate the functions

$$\rho'(t,x') = \rho(t,x),$$

$$\underset{\sim}{t}'(t,x') = Q(t)\underset{\sim}{t}(t,x)Q(t),$$

$$\underset{\sim}{n}' = Q\underset{\sim}{n},$$

$$q'(t,x') = Q(t)q(t,x),$$

$$r'(t,x') = r(t,x),$$

$$e'(t,x') = e(t,x),$$

and finally define b' through

$$\rho' b' - \rho' a' = Q(t)(\rho b - \rho a).$$

The last condition asserts that $b - a$ should transform as a vector. Thus b' will include b plus fictitious forces due to an accelerating rigid frame[†].

We <u>assume that the new motion with the corresponding primed quantities satisfies balance of energy</u>.

<u>Theorem</u> (Green and Rivlin [101])

Under the assumptions just described, the local forms of conservation of mass, balance of momentum, balance of moment of momentum, and balance of energy all hold. Conversely, if these balance principles all hold, then balance of energy is invariant under time-dependent isometries of $\mathbb{R}^3$.

<u>Proof</u>. It suffices to prove the first part at $t_0 = 0$. First, let $\psi(t,x) = tc + x$, $c = \text{const.}$, so $v' = v + c$. Thus, by the localized energy balance for the primed quantities at $t = 0$ (where $x = x'$),

$$\rho'(\dot{e}' + \langle v', a'\rangle) + (\dot{\rho}' + \rho' \operatorname{div} v)(e' + \tfrac{1}{2}\langle v', v'\rangle)$$

$$= \rho' r' + \rho'\langle b', v'\rangle - \operatorname{div} q' + \operatorname{div}(t' v'),$$

where

$$\rho' = \rho, \quad \dot{e}' = \dot{e}, \quad v' = v + c, \quad \dot{\rho}' = \dot{\rho}, \quad \operatorname{div} v' = \operatorname{div} v, \quad r' = r,$$

[†]One can legitimately object to this definition of b'. However this can be overcome by lumping the terms from the kinetic energy and the external forces and working with the "noninertial" part of the forces from the start. See also Noll [202] where forces are derived from the more primitive concept of power.

$$\rho' b' - \rho' a' = \rho(b - a), \quad \operatorname{div} q' = \operatorname{div} q, \quad \text{and} \quad t' = t.$$

Thus, for the unprimed quantities we get the identity

$$\langle \rho a - \rho b - \operatorname{div} \underset{\sim}{t}, c \rangle + (\dot{\rho} + \rho \operatorname{div} v)\langle v, c \rangle + \tfrac{1}{2}(\dot{\rho} + \rho \operatorname{div} v)\langle c, c \rangle = 0$$

for all c. Letting $c = \lambda u$ where u is a unit vector, differentiating twice with respect to λ and setting $\lambda = 0$ gives $\dot{\rho} + \rho \operatorname{div} v = 0$, i.e., conservation of mass. Inserting this back into the identity gives, since c is arbitrary, $\rho a - \rho b - \operatorname{div} t = 0$; i.e., balance of momentum.

Next, let $\psi(t,x) = Q(t)x$ with $Q(0) = \text{Identity}$. Then at $t = 0$, $v' = \underset{\sim}{\Omega} x + v$ where $\underset{\sim}{\Omega} = \dot{Q}(0)$, a skew symmetric matrix. Comparing energy balance for primed and unprimed quantities, using $\operatorname{div} v' = \operatorname{div} v$, conservation of mass and balance of momentum now merely yields

$$\underset{\sim}{t} : \underset{\sim}{\Omega} = 0.$$

In doing this, note that

$$\begin{aligned} \operatorname{div}(t_{ij}\Omega_{jk}x^k) &= \frac{\partial}{\partial x^i}(t_{ij}\Omega_{jk}x^k) \\ &= (\operatorname{div} \underset{\sim}{t}) \cdot (\underset{\sim}{\Omega} x) + t_{ij}\Omega_{jk}\delta_i^k \\ &= \operatorname{div} \underset{\sim}{t} \cdot (\underset{\sim}{\Omega} x) + t_{ij}\Omega_{ji} . \end{aligned}$$

But Ω_{ij} is an arbitrary skew symmetric matrix. Thus t_{ij} must be symmetric; i.e., balance of moment of momentum must hold.

The converse assertion can be easily checked, the details of which we leave to the reader. □

In Sections 9 and 10 we shall see how the same conclusions can be reached if we start with elasticity as a classical Lagrangian field theory, impose invariance under one parameter groups of spatial translations and rotations, and use Noether's theorem. The two approaches are almost the same thing but done from different points of view.

Covariant classical field theory (and general relativity) teaches us that the stress tensor should be related to the Hamiltonian or energy density $\mathcal{H}$

$$\underset{\sim}{t} = \frac{\partial \mathcal{H}}{\partial g},$$

where g is the metric on the space. See, for example Hawking and Ellis [119]. Except for works in relativistic elasticity, this formula has not yet penetrated the classical literature.† We shall now show how this formula is in fact needed to make the Green-Rivlin formulation fully covariant (and in particular, localizable and not dependent on inertial frames).

One can see that something else besides balance of energy is needed, for if it holds under all superposed motions, not necessarily rigid body motions, then the stress tensor $\underset{\sim}{t}$ would be identically zero.*

We shall now present a covariant version of the Green-Rivlin theorem. Although it does not deal with any constitutive hypotheses per se, the ideas are related to a covariant version of constitutive theory described in Section 3.

Suppose we have a motion ϕ_t of M in N and associated with it we have functions e, $\underset{\sim}{t}, \cdots$ as above. (We again assume $\tau = \langle \underset{\sim}{t}, n \rangle$ and $h = \langle q, n \rangle$ for simplicity, although these facts may be deduced.)

---

†As we shall see, it is closely related to the classical formula $T = \rho_0 \partial W/\partial F$ or $P = 2\rho_0 \partial W/\partial C$.

*However, this works if one assumes a localized form of energy balance and superposes infinitesimal isometries. See Hughes and Marsden [130].

If we fix our metric g (say the standard Euclidean metric) on N, nevertheless g will be represented differently in different coordinate systems and the coordinate representations of e, $\underset{\sim}{t}$,··· may reflect this. (We shall explore this idea in Section 3.) However, we contemplate now something more drastic, as covariance suggests we should. Namely, think of the same motion for <u>different</u> choices of metric g on N.

Changing the metric on N of course changes (pointwise) our units of measurement but more dramatically, it changes the accelerations of particles. For instance, in $\mathbb{R}^3$ with a non-Euclidean metric, particles moving uniformly on a straight line may be accelerating.

Such a change will, therefore, change our functions e, $\underset{\sim}{t}$,···. It must, because the very equations of motion are no longer the same since forces and accelerations are modified.

We postulate three things:

<u>Assumption 1</u> For a given motion ϕ_t and associated functions e, $\underset{\sim}{t}$,···, energy balance is satisfied. If the metric g on N is changed, the functions e, $\underset{\sim}{t}$,··· may change, but balance of energy with the new quantities will be maintained. The dependence on the metric is denoted e(t,x,g) etc.

<u>Assumption 2</u> For any superposed motion ξ_t on N define the primed quantities as above, except using the metric ξ_t^*g, i.e.,

$$e'(t,x',g) = e(t,x,\xi_t^*g)$$

etc. Then the primed quantities also satisfy balance of energy in N with respect to the metric g.
(The idea is schematically indicated in figure 2.)

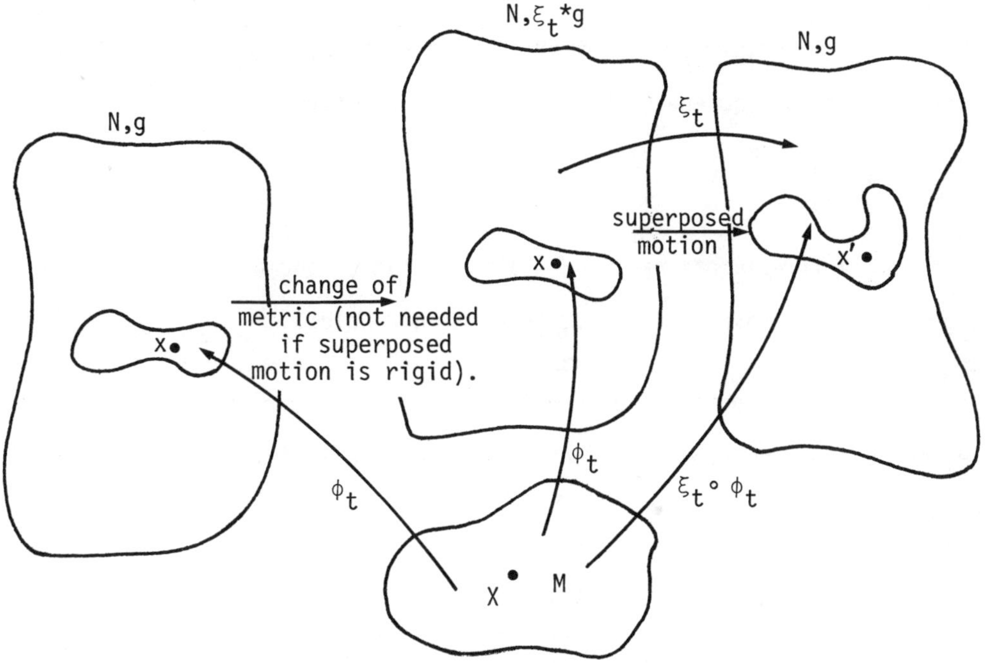

Fig. 2.

If these two assumptions are made, we say that energy balance is covariant. Notice that in $\mathbb{R}^3$, if ψ_t is a rigid motion, then Assumption 2 reduces to the previous assumption of Green and Rivlin, since ψ_t leaves the metric unchanged.

Assumption 3 The dependence of e on g is local; i.e., e depends (differentiably) on the point values of g and its derivatives up to some finite order.

The third assumption is of a type found in constitutive theory and is explained in greater detail in the next section. However, it is to be stressed that we are not making constitutive assumptions here; no special dependence of e on ϕ is assumed. This will all be treated together in the next section.

Theorem If Assumptions 1, 2 and 3 hold, then e depends only on the point values of g. Furthermore, conservation of mass, balance of momentum and moment of momentum and energy hold in their local forms and

$$\underset{\sim}{t} = 2\rho \frac{\partial e}{\partial g} .$$

Conversely if Assumptions 1 and 3 and all these conclusions hold, then energy balance is covariant, i.e., Assumption 2 holds.

Proof. Referring to the Green-Rivlin argument above, we had $(e')^{\cdot} = \dot{e}'$. Here, because of the dependence on the metric, the definition of material time derivative yields

$$(e')^{\cdot} = \dot{e} + \frac{\partial e}{\partial g} : \mathcal{L}_w g + \text{h.o.t.},$$

where $\frac{\partial e}{\partial g}$ stands for the derivative of e with respect to the point values of g and h.o.t. contains analogous terms involving derivatives of e with respect to point values of the derivatives of g (e.g., the connection of g, curvature, etc.).

In that argument we also pick up an extra term

$$\underset{\sim}{t}:k = \tfrac{1}{2}\underset{\sim}{t}:\mathcal{L}_w g$$

as follows.

We notice, first of all, that

$$v'_t = w_t + \xi_{t*} v_t$$

where w_t is the velocity field of ξ_t, as usual. This follows from the definition $\phi'_t(X) = \psi_t(\phi_t(X))$. Localized balance of energy now reads:

$$\rho(\dot{e} + \langle v,a \rangle) + (\dot{\rho} + \rho \operatorname{div} v)(e + \tfrac{1}{2}\langle v,v \rangle) = \rho r + \rho\langle b,v \rangle - \operatorname{div} q + \operatorname{div}(\underset{\sim}{t} \cdot v). \tag{1}$$

By hypothesis, this also holds for the primed quantities. Now write:

$$\operatorname{div}(\underset{\sim}{t}' \cdot v') = \langle \operatorname{div} \underset{\sim}{t}', v' \rangle + \underset{\sim}{t}' : d' - \underset{\sim}{t}' : \omega',$$

where $\omega'_{ab} = \tfrac{1}{2}(v'_{a|b} - v'_{b|a})$.

Thus, (1) for the primed quantities reads:

$$\begin{aligned} 0 = {} & \rho'([e']^{\cdot} - r') + \operatorname{div} q' - \underset{\sim}{t}' : d' \\ & + \langle v', \rho'(a' - b') - \operatorname{div} \underset{\sim}{t}' \rangle \\ & + (\dot{\rho}' + \rho' \operatorname{div} v')(e' + \tfrac{1}{2}\langle v', v' \rangle) \\ & + \underset{\sim}{t}' : \omega' . \end{aligned} \tag{2}$$

At t,x, chosen so that $\xi_t(x) = x$, some simplification of (2), using $v' = v + w$ and (1), yields

$$\begin{aligned} 0 = {} & \rho\left(\frac{\partial e}{\partial g} : \mathcal{L}_w g + \text{h.o.t}\right) \\ & + \{\langle w, \rho(a - b) - \operatorname{div} \underset{\sim}{t} + (\dot{\rho} + \rho \operatorname{div} v)v \rangle\} \\ & + \{\tfrac{1}{2}(\dot{\rho} + \rho \operatorname{div} v)\langle w,w \rangle\} \\ & + \underset{\sim}{t} : \omega - \underset{\sim}{t} : k, \end{aligned} \tag{3}$$

where we have written

$$\omega_{ab} = \tfrac{1}{2}(w_{a|b} - w_{b|a}), \qquad k_{ab} = \tfrac{1}{2}(w_{a|b} + w_{b|a})$$

and used the identity

$$d'_{ab} = d_{ab} + k_{ab}.$$

In (3), h.o.t. is the only term involving second or higher derivatives of w and g. Since these are arbitrary, one concludes that $\text{h.o.t} = 0$ and that e depends only on the point values of g. Using the pointwise arbitrariness of ω, k, and w, the rest of the conclusions follow as in the Green-Rivlin theorem.

The converse may be proved by a similar argument. (Note that (3) will hold for the primed quantities without necessarily assuming $\psi_t = \text{id}$ since we can transform the unprimed quantities - relative to the metric $\xi_t^* g$.) □

Notice that assumption 1 is analogous to a common assumption in constitutive theory regarding the admission of a whole class of processes. Indeed, if the metric changes, then so do the accelerations and hence the process itself changes.

The following is a list of miscellaneous comments on the above result.

1 The theorem includes thermo-elasticity as well as hyper-elasticity. When constitutive theory is introduced, $\frac{\partial e}{\partial g}$ must be interpreted as a derivative at constant entropy η so it can be equated with $\frac{\partial \psi}{\partial g}$ where ψ is the free energy, with the derivative taken at constant temperature θ. In other words, in the theorem, the basic fields are (ϕ,η) and not (ϕ,θ); we will review this fact in terms of Legendre transformations in the next section.

2 The theorem can be modified to include rate or memory effects. Indeed, if ψ_t is not rigid, it is natural to expect it not to transform $\underset{\sim}{t}$ as a tensor, if $\underset{\sim}{t}$ depends on rates of shearing for example. Indeed one can use this notion to <u>define</u> a fluid etc. Notice that this is done outside of constitutive theory. All meterials can be included by assuming a <u>general</u> transformation law for $\underset{\sim}{t}$. Specializing to

(a) $\underset{\sim}{t}' = \xi_* \underset{\sim}{t}$ yields elasticity,

(b) $\underset{\sim}{t}' = \xi_* \underset{\sim}{t} + vk$ yields elastic fluids, etc. (Here $k_{ab} = \frac{1}{2}(w_{a|b} + w_{b|a})$ and w is the velocity of the superposed motion as above.)

3 In Lianis and Rivlin [173] the Green-Rivlin theorem is generalized to <u>special</u> relativity. For <u>general</u> relativity one requires ideas introduced here. (See also Carter and Quintana [36].)

4 There appear to be some links between the ideas presented here and the work of Noll [202]. So far these links are unexplored.

5 The covariance of energy balance allows one to proceed <u>directly</u> from it to the weak form of the equations <u>without</u> first passing to the localized form as is usually done. This is possible because we allow the superposed motion ψ_t to be arbitrary, not necessarily rigid. (This can also be done from the viewpoint of Lagrangian field theory; cf. Sections 9 and 10.) This remark is inspired by similar comments of S. Antman in a different context.

We conclude with some standard results on the entropy production inequality, since we shall need them in the next section.

Let $\phi(t,X)$ be a regular motion of a simple body $M \subset N$. In addition to functions $\rho(t,x)$, $v(t,x)$, $r(t,x)$ and $h(t,s)$, introduced earlier, assume there are functions $\eta(t,x)$, the specific (per unit mass) <u>entropy</u> and $\theta(t,x) > 0$, the <u>absolute temperature</u>.

<u>Definition</u> These functions are said to obey the <u>entropy production inequality</u> or the <u>Clausius-Duhem inequality</u> if, for all (nice) $U \subset M$, we have

$$\frac{d}{dt}\int_{\phi_t(U)} \rho\eta \, dv \geq \int_{\phi_t(U)} \frac{\rho r}{\theta} dv + \int_{\partial\phi_t(U)} \frac{h}{\theta} da.$$

Let us assume, as above, that $h(t,x,n) = -\langle q(t,x), n\rangle$ and conservation of mass holds. Then the entropy production inequality localizes (by the Master balance inequality) to:

$$\rho\dot{\eta} \geqslant \frac{\rho r}{\theta} - \operatorname{div}\left(\frac{q}{\theta}\right).$$

For the material form, we let

$$\Theta(t,X) = \theta(t,x), \quad N(t,X) = \eta(t,x), \quad R(t,X) = r(t,x), \quad \rho_0(X) = \rho(0,X),$$

$$Q(t,X) = JF^{-1}q(t,x). \quad \text{(Piola transform).}$$

The material form of the entropy production inequality is

$$\frac{d}{dt}\int_U \rho_0 N\, dV \geqslant \int_U \rho_0 R\, dV + \int_{\partial U} \frac{Q.N}{\Theta} dA,$$

or, in localized form,

$$\rho_0 \frac{\partial N}{\partial t} \geqslant \rho_0 \frac{R}{\Theta} - \operatorname{DIV}\left(\frac{Q}{\Theta}\right).$$

It is quite convenient to recast the entropy production inequality into a slightly different form by bringing in the free energy $\psi(t,x)$.

The free energy is defined by

$$\psi = e - \theta\eta,$$

or, materially,

$$\Psi = E - \Theta N.$$

Proposition Assume conservation of mass, balance of momentum, moment of momentum, energy and the entropy production inequality hold. Then

$$\rho(\eta\dot{\theta} + \dot{\psi}) - \underset{\sim}{t}:d + \frac{1}{\theta}\langle q, \nabla\theta \rangle \leqslant 0,$$

or, materially,

$$\rho_0\left(N\frac{\partial\Theta}{\partial t}+\frac{\partial\Psi}{\partial t}\right) - T : \frac{\partial F}{\partial t} + \frac{1}{\Theta}\langle Q, \mathrm{GRAD}\,\Theta\rangle \leq 0.$$

This is called the spatial (resp. material) <u>reduced dissipation inequality</u>.

<u>Proof</u>. From the definition of ψ,

$$\dot{\psi} = \dot{e} - \dot{\theta}\eta - \theta\dot{\eta}, \quad \text{i.e.,} \quad \theta\dot{\eta} = \dot{e} - \dot{\theta}\eta - \dot{\psi}\ .$$

Combining this with the localized entropy production inequality gives

$$\rho(\dot{e} - \dot{\theta}\eta - \dot{\psi}) \geq \rho r - \operatorname{div} q + \frac{1}{\theta}\langle q, \nabla\theta\rangle\ .$$

From localized balance of energy (and the other balance laws), we have

$$\rho\dot{e} = \rho r - \operatorname{div} q + \underset{\sim}{t}:d,$$

which on substitution gives the spatial result. The material form is proved the same way. □

If we apply the earlier covariance argument to the entropy production inequality straight away, we would get $\frac{\partial\eta}{\partial g} = 0$, i.e., the entropy should not change when g changes. This then merely reaffirms our earlier statement that for thermo-elasticity, in spatial covariance, η and g should be regarded as independent variables.

On the other hand, if we take the reduced dissipation inequality, apply the covariance assumption, and regard θ, g as independent (i.e., $\frac{\partial\theta}{\partial g} = 0$), then we find that ψ depends only on point values of g and that

$$2\rho\frac{\partial\psi}{\partial g} = \underset{\sim}{t}.$$

Therefore, we are seeing a manifestation of the usual thermodynamic relationship $\eta = -\partial\psi/\partial\theta$ on grounds of covariance alone. With a constitutive assumption here, this may be derived.†

In other words, standard balance assumptions concerning a single motion in all possible geometries may be used in place of assumptions concerning all conceivable motions in a fixed Euclidean geometry, to derive the thermodynamic relations.

Finally, we notice that in the approach using covariance, we do not need to postulate the Clausius-Duhem inequality (not everyone is willing to do so) but only that the entropy production

$$\gamma = \rho\dot{\eta} - \frac{\rho r}{\theta} + \operatorname{div}\left(\frac{q}{\theta}\right)$$

$$= \frac{-1}{\theta}\{\rho(\eta\dot{\theta} + \dot{\psi}) - \underset{\sim}{t}:d + \frac{1}{\theta}\langle q, \nabla\theta\rangle\}$$

satisfies a covariance assumption. (We thank M. Gurtin for pointing this out.)

3 CONSTITUTIVE THEORY

We shall begin by reviewing the standard approach to constitutive theory following Truesdell and Noll [248] and Gurtin [107]. At first we stick to thermoelastic solids for simplicity, i.e., we shall ignore rate and memory effects for the moment. Following this we shall mention an alternative approach using the covariance ideas from the preceding section.

One of our main goals is to explain how to formulate material frame indifference (MFI) in a covariant way. Our approach is entirely equivalent to the usual one in Euclidean space, but it is formulated in a way which makes sense on any manifold and does not refer to rigid motions. This complements our covariant version of the Green-Rivlin argument given in Section 2.

---

†As earlier, rate of memory effects require a more general assumption on the transformation law of $\underset{\sim}{t}$.

By allowing dependence on the metric we get a more symmetric relationship between the material and spatial pictures, as shown in Figure 3. This leads to a better understanding of the relationship between <u>spatial covariance</u> and <u>material symmetries</u>, a topic taken up again from a Hamiltonian point of view in Sections 9 and 10.

Finally, we summarize some of the important definitions and relationships concerning the elasticity tensor for later use.

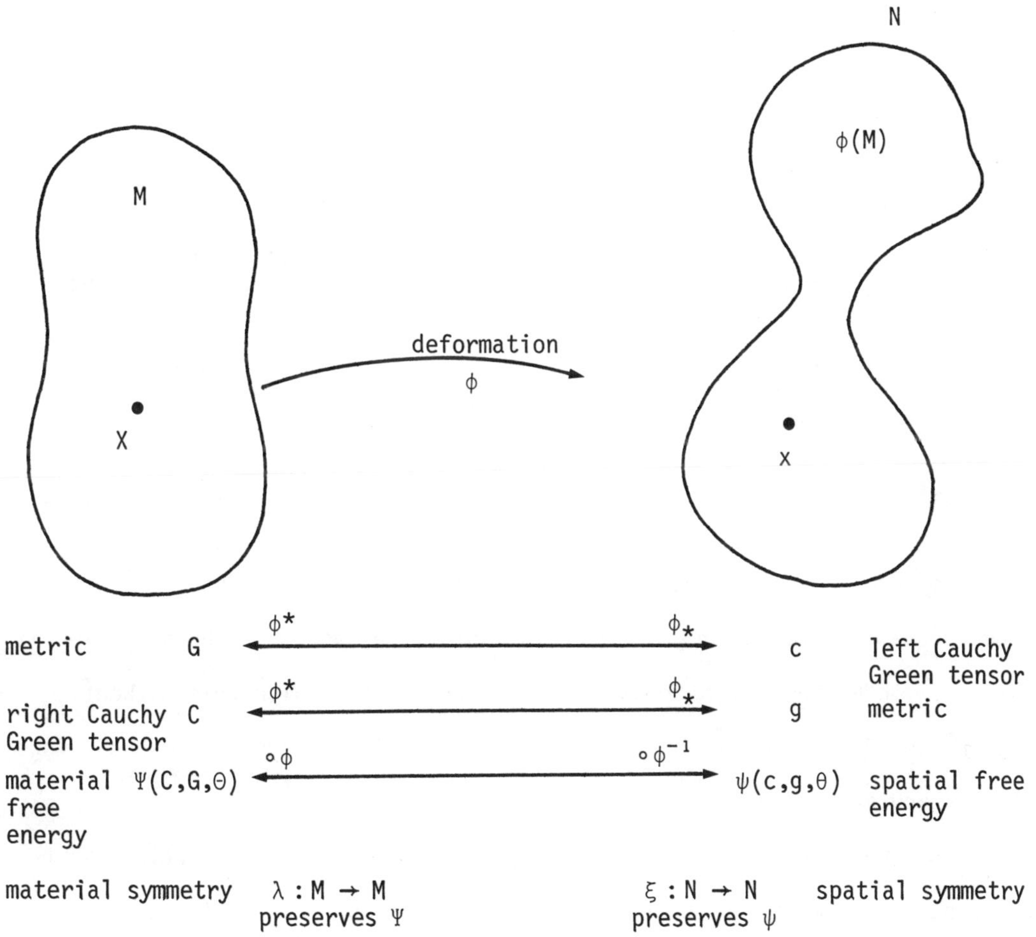

Fig. 3.

Constitutive theory gives functional form to the stress tensor, free energy and heat flux vector in terms of the motion and temperature (or the motion and entropy). Such functional forms are needed not only to distinguish broad classes of materials or to single out specific ones, but to make the equations of motion formally well posed (in the naive sense that we have as many equations as unknowns).

We begin with the traditional approach (trivially placed in a manifold context) with everything done in the material picture. From the previous section, we have:

(i) $\rho_0 = \rho J$ (conservation of mass)

(ii) $\rho_0 \frac{\partial \underset{\sim}{V}}{\partial t} = \mathrm{DIV}(\underset{\sim}{P}\underset{\sim}{F}^T) + \rho_0 \underset{\sim}{B}$ (balance of momentum)

(iii) $\underset{\sim}{P} = \underset{\sim}{P}^T$ (symmetry; balance of moment of momentum)

(iv) $\rho_0 \frac{\partial E}{\partial t} + \mathrm{DIV}\,\underset{\sim}{Q} = \rho_0 R + \underset{\sim}{P}:\underset{\sim}{D}$ (balance of energy)

(v) $\rho_0 N \frac{\partial \Theta}{\partial t} + \frac{\partial \Psi}{\partial t} - \underset{\sim}{P}:\underset{\sim}{D} + \frac{1}{\Theta}\langle Q, \mathrm{GRAD}\,\Theta\rangle \leqslant 0$

(reduced dissipation inequality).

These equations are formally ill posed in the sense that there are not enough equations to determine the evolution of the system. The situation is analogous to Newton's second law $m\ddot{x} = F$; one cannot solve this without saying how F depends on x and $\dot{x}$. Doing so is tantamount to specifying the particular system under study.

Often, one regards the motion $\phi(t,X)$ and the temperature $\Theta(t,X)$ as the unknowns and attempts to solve for them from equations (ii), (iv) and their initial (Cauchy) data. If ϕ is known and ρ_0 is given, (i) determines ρ, so we can eliminate condition (i). We often regard $\underset{\sim}{B}$ and R as given externally. Thus, if we are going to determine ϕ and Θ, we must specify $\underset{\sim}{P}$, $\underset{\sim}{Q}$ and E as functions of ϕ and Θ. Since E is given by $E = \Psi + N\Theta$, we might also ask for N and Ψ to be functions of ϕ and Θ. Of course, (iii) and (v) are required to hold. In the traditional approach the reduced dissipation inequality and the concept of material frame indifference play a central role.

If we eliminate rate and memory effects for simplicity, a (thermo-elastic) constitutive function for, say, P, the second Piola-Kirchhoff stress tensor, is a mapping

$$\hat{P}: \mathscr{K} \times \mathcal{R} \longrightarrow S_2(M),$$

where $\mathscr{K}$ is the space of configurations $\phi: M \longrightarrow N$, $\mathcal{R}$ is the space of temperature fields; i.e., of positive functions $\Theta: M \longrightarrow \mathbb{R}$, and $S_2(M)$ is the space of symmetric two-tensor fields on M. It is tacitly assumed that in appropriate function space topologies, $\hat{P}$ is a differentiable function in the Fréchet sense.

The second Piola-Kirchhoff stress tensor associated with a motion $\phi(t,X)$ and a temperature field $\Theta(t,X)$ is then

$$P(t,X) = \hat{P}(\phi_t,\Theta_t)(X).$$

Constitutive functions for Q, N and Ψ are defined similarly. For instance, we have

$$\hat{\Psi}: \mathscr{K} \times \mathcal{R} \longrightarrow \mathscr{F}(M),$$

where $\mathscr{F}(M)$ is the space of scalar functions on M.

Definition A constitutive function for thermo-elasticity, say

$$\hat{\Psi}: \mathscr{K} \times \mathscr{R} \longrightarrow \mathscr{F}(M)$$

is called *local* if for any open set $U \subset M$ and $\phi_1, \phi_2 \in \mathscr{K}$ with $\phi_1 = \phi_2$ on U and $\Theta_1, \Theta_2 \in \mathscr{R}$ with $\Theta_1 = \Theta_2$ on U, we have

$$\hat{\Psi}(\phi_1, \Theta_1)(X) = \Psi(\phi_2, \Theta_2)(X)$$

for all $X \in U$.

Notice that if $\hat{\Psi}$ depends on the point values of ϕ and Θ and their derivatives up to order, say, k, then $\hat{\Psi}$ is local. This is merely because knowledge of a mapping on an open set entails a knowledge of all its derivatives on that set. A $\hat{\Psi}$ of this form is called a (non-linear) *differential operator*.

While it is trivial that a differential operator is local, the converse is not so elementary. For linear operators, this is due to Peetre(Math. Scand. 7(1959)211-218). For non-linear operators, it was proved, under some annoying technical assumptions (namely that the linearizations are local - and hence are differential operators - of bounded order) by Dombrowski (Nachr. Akad. Wiss. Gott. Kl. II (1966)19-43). A satisfactory general theorem is not yet known, but it seems plausible, perhaps using methods of Terng [238].

Fortunately, a basic observation of Gurtin [107], given below, enables us to bypass this point for most considerations.

The idea of using locality as a basic postulate is due to Noll [201]. It must be emphasized that one may wish to add on *nonlocal* constraints, such as incompressibility (See Section 10).

First Axiom of Constitutive Theory Constitutive functions for thermo-elasticity are assumed to be local.

We next invoke the entropy production inequality in a rather strong way.

Namely we assume it holds for all (regular) motions of the body. The momentum balance and energy balance are not taken into account, because any motion is consistent with them for a suitable choice of body force $\underset{\sim}{B}$ and heat source R, i.e., balance of momentum and energy define what $\underset{\sim}{B}$ and R have to be. This is not unreasonable since we are supposed to be able to allow any choice of $\underset{\sim}{B}$ and R.

Second Axiom of Constitutive Theory For any (regular) motion of M, constitutive functions for thermo-elasticity are assumed to satisfy the entropy production inequality:

$$\rho_0\left(\hat{N}\frac{\partial\Theta}{\partial t}+\frac{\partial\hat{\Psi}}{\partial t}\right) - \hat{T}:\frac{\partial F}{\partial t} + \frac{1}{\Theta}\langle \hat{Q}, \mathrm{GRAD}\,\Theta\rangle \leqslant 0.$$

Theorem (Coleman and Noll [56]) Suppose the axioms of locality and entropy production hold. Then $\hat{\Psi}$ depends only on the variables X, F and Θ. Moreover, we have

$$\hat{N} = -\frac{\partial\hat{\Psi}}{\partial\Theta} \quad \text{and} \quad \hat{T} = \rho_0\frac{\partial\hat{\Psi}}{\partial F}\cdot g^{\sharp}, \quad \text{i.e.,} \quad \hat{T}^{Aa} = \rho_0\frac{\partial\hat{\Psi}}{\partial F^b{}_A}g^{ab}$$

and the entropy production inequality reduces to

$$\langle Q, \mathrm{GRAD}\,\Theta\rangle \leqslant 0.$$

Proof. (Gurtin [107]) Write

$$\frac{\partial\hat{\Psi}}{\partial t} = D_{\phi}\hat{\Psi}\cdot V + D_{\Theta}\hat{\Psi}\cdot\dot{\Theta},$$

by the chain rule. Here, $D_{\phi}\hat{\Psi}$ is the linearization of Ψ in the Fréchet sense ($\hat{\Psi}$ depends on the function ϕ and $D_{\phi}\hat{\Psi}$ may be a differential operator). Thus the entropy production inequality reads

$$\rho_0(\hat{N}\dot{\Theta} + D_\Theta\hat{\Psi}\cdot\dot{\Theta}) + (\rho_0 D_\phi\hat{\Psi}\cdot V - \hat{T}:\dot{F}) + \frac{1}{\Theta}\langle \hat{Q}, \mathrm{GRAD}\,\Theta\rangle \leqslant 0.$$

Since this holds for all processes, the first two terms vanish identically. For instance, fixing V, Θ, ϕ, $\dot{F}$ at an instant, we can choose $\dot{\Theta}$ arbitrarily. If $\hat{N} + D_\Theta\hat{\Psi}$ were not zero, the inequality could be violated by choosing $\dot{\Theta}$ appropriately; for example, by replacing $\dot{\Theta}$ by $\alpha\dot{\Theta}$ where α is a constant of arbitrary sign.

Thus we have, identically,

$$\hat{N}\dot{\Theta} + D_\Theta\hat{\Psi}\cdot\dot{\Theta} = 0,$$

and

$$\rho_0 D_\phi\hat{\Psi}\cdot V = \hat{T}:\dot{F}.$$

Consider the second equality, and fix $X \in M$ and Θ. Let ϕ_0 and ϕ_1 be two configurations with $F_0(X) = F_1(X)$. Let (in a chart on N) $\phi(t,Y) = \phi_0(Y) + t(\phi_1(Y) - \phi_0(Y))$ define ϕ for Y near X (and arbitrary outside a neighborhood of X). Then $F(t,X) = F_0(X)$ is constant in time, and so $\dot{F}(X) = 0$. Thus, from the second equality,

$$\frac{d}{dt}\rho_0(X)\hat{\Psi}(\phi_t,\Theta)(X) = \rho_0(X)(D_\phi\hat{\Psi}\cdot V)(X) = \hat{T}:\dot{F}(X) = 0,$$

and so

$$\rho_0(X)\hat{\Psi}(\phi_0,\Theta)(X) = \rho_0(X)\hat{\Psi}(\phi_1,\Theta)(X).$$

Thus, $\hat{\Psi}$ can depend on ϕ only through the pointwise values of F. Similarly, it can depend only on the pointwise values of Θ, so the theorem follows. □

Notice that locality is needed so that we can use the formula for $\phi(t,Y)$ (which may be regular only for Y near X) only near X and get an answer independent of how ϕ may be extended to a global regular configuration.

Another noteworthy remark is that the relationship $\hat{T} = \rho_0(\partial\hat{\Psi}/\partial F)\cdot g^\sharp$ can be derived from the first law (energy balance) if only the first axiom of constitutive theory is assumed and Ψ is a function of X, F and Θ.

We cannot conclude anything about the dependence of $\hat{Q}$ on ϕ and Θ. It could conceivably depend on many derivatives. It is usually assumed (in an apparently ad hoc manner) that $\hat{Q}$ depends on X and the point values of F, Θ and $\mathrm{GRAD}\,\Theta$. (A "grade (1,1)" material.)

Two standard consequences of the above theorem follow:

<u>Proposition</u> Q vanishes when its argument $\mathrm{GRAD}\,\Theta$ vanishes.

<u>Proof</u>. Fixing all other arguments, let $f(\alpha) = \langle Q(\alpha\,\mathrm{GRAD}\,\Theta), \mathrm{GRAD}\,\Theta\rangle$. Thus $\alpha f(\alpha) \leqslant 0$, so f changes sign at $\alpha = 0$. Since f is continuous, $f(0) = 0$. Thus $\langle Q(0), \mathrm{GRAD}\,\Theta\rangle = 0$, so $Q(0) = 0$. □

Thus for $\mathrm{GRAD}\,\Theta$ small, by Taylor's Theorem, $\hat{Q}$ is well approximated by a matrix which is negative semi-definite times $\mathrm{GRAD}\,\Theta$. More precisely, again fixing all arguments but $\mathrm{GRAD}\,\Theta$,

$$\hat{Q}(\mathrm{GRAD}\,\Theta) - A\cdot\mathrm{GRAD}\,\Theta,$$

where

$$A = \int_0^1 \frac{\partial\hat{Q}}{\partial(\mathrm{GRAD}\,\Theta)}(s\,\mathrm{GRAD}\,\Theta)ds$$

by the fundamental theorem of calculus. The inequality $\langle\hat{Q}, \mathrm{GRAD}\,\Theta\rangle \leqslant 0$ means A is negative semi-definite, i.e., dissipative. If A were assumed constant one would recover the Fourier law of heat conduction.

<u>Proposition</u> Balance of energy,

$$\rho_0\frac{\partial E}{\partial t} + \mathrm{DIV}\,Q = T : \frac{\partial F}{\partial t} + \rho_0 R,$$

may be rewritten (on a motion and temperature field) as

$$\rho_0\Theta\frac{\partial}{\partial t}\hat{N} + \mathrm{DIV}\,\hat{Q} = \rho_0 R.$$

Proof. We have $\hat{E} = \hat{\Psi} + \hat{N}\Theta$, so

$$\frac{\partial\hat{E}}{\partial t} = \frac{\partial\hat{\Psi}}{\partial t} + \frac{\partial\hat{N}}{\partial t}\Theta + \hat{N}\frac{\partial\Theta}{\partial t}$$

$$= \frac{\partial\hat{\Psi}}{\partial F}:\frac{\partial F}{\partial t} + \frac{\partial\hat{\Psi}}{\partial\Theta}\frac{\partial\Theta}{\partial t} + \frac{\partial\hat{N}}{\partial t}\Theta + \hat{N}\frac{\partial\Theta}{\partial t}.$$

On using $\rho_0\frac{\partial\hat{\Psi}}{\partial F}:\frac{\partial F}{\partial t} = \hat{T}:\frac{\partial F}{\partial t}$ and $\frac{\partial\hat{\Psi}}{\partial\Theta} = -\hat{N}$ and substituting $\frac{\partial E}{\partial t}$ in the balance of energy, these terms cancel out. □

Example (Rigid Heat Conductor) For the rigid heat conductor we make the assumption that the motion is fixed, say ϕ = identity, for all time. We also assume that $\hat{Q}$ depends only on X, Θ and GRAD Θ. Then the evolution of Θ in time is determined from balance of energy, namely,

$$\rho_0\Theta\frac{\partial\hat{N}}{\partial t} + \mathrm{DIV}\,\hat{Q} = \rho_0 R.$$

Since $\hat{N}$ depends only on X and Θ, we get

$$\left(\rho_0\Theta\frac{\partial\hat{N}}{\partial\Theta}\right)\frac{\partial\Theta}{\partial t} = -(\hat{Q}^A)_{|A} + \rho_0 R$$

$$= -\left(\frac{\partial\hat{Q}^A}{\partial\Theta_{|B}}\right)\Theta_{|B|A} + \frac{\partial\hat{Q}^A}{\partial\Theta}\Theta_{|A} + \rho_0 R.$$

As was observed above, the matrix $\partial\hat{Q}^A/\partial\Theta_{|B}$ is negative semi-definite. We also assume the scalar function $\partial\hat{N}/\partial\Theta = -(\partial^2\hat{\Psi}/\partial\Theta^2)$ (= specific heat at constant volume) is positive, so the equation is formally parabolic. It is the general form of a nonlinear heat equation. Notice, finally, that positivity of $\partial^2\hat{\Psi}/\partial\Theta^2$ means $\hat{\Psi}$ is a convex function of Θ.

The Clausius-Duhem inequality almost implies that the operator $\Theta \longmapsto (\hat{Q}^A)_{|A}$ is monotone (see the lectures of A. Pazy in the next volume).

Even granting this, and positivity of $\partial\hat{N}/\partial\Theta$, it is not yet known if the full rigid heat conductor has an associated global existence and uniqueness theorem. (cf. Tartar [237]). As we shall discuss in the linear case in Section 7, if one assumes that the equations are well posed (in the sense of defining a continuous dynamical system on an appropriate space - See Section 9) then positivity of $\partial N/\partial\Theta$ will probably follow. Correspondingly, for well posedness of the equations of motion for ϕ, an assumption of strong ellipticity is needed. This will be discussed in due course.

Now we turn to the final constitutive axiom, namely material frame indifference.

<u>The Third Axiom of Constitutive Theory (Material Frame Indifference)</u>† Let $\hat{\Psi}$ be a thermo-elastic constitutive function satisfying the above axioms, so that $\hat{\Psi}$ is a function of X, (and the point values of) F and Θ. Assume that if $\xi : N \longrightarrow N$ is a regular map taking x to x' and $T\xi$ is an isometry from T_xN to $T_{x'}N$, then

$$\hat{\Psi}(X,F,\Theta) = \hat{\Psi}(X,F',\Theta),$$

where $F : T_XM \longrightarrow T_xN$, $F' : T_XM \longrightarrow T_{x'}N$ and $F' = T\xi\cdot F$. (See Fig. 4.)

Stated loosely, this axiom means that our constitutive functions are invariant under rotations of the ambient space N in which our body moves. One sometimes says that $T\xi$ 'rotates observer frames'.

†While this axiom is widely accepted, some objections have been raised. See, for instance, Muller [192]. (His objections rely on relativistic thermal effects, a topic which itself is objectionable!).

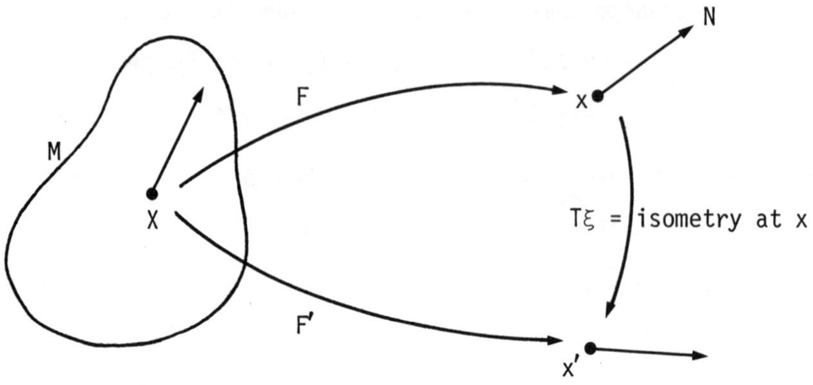

Fig. 4.

Theorem Let $\hat{\Psi}$ satisfy the above axioms. Then $\hat{\Psi}$ is a function only of X, C and Θ. By abuse of notation we shall write

$$\hat{\Psi}(X,C,\Theta).$$

Remarks 1. Equally well, we could say $\hat{\Psi}$ is a function only of X, Θ and the right stretch tensor $U = C^{\frac{1}{2}}$.

2. By abuse of notation we shall write C for both C and $C^{\sharp}$, i.e., C with its indices raised by G, when there is little danger of confusion.

Proof. Suppose $F_1: T_X M \longrightarrow T_{x_1} N$ and $F_2: T_X M \longrightarrow T_{x_2} N$ and $F_1^T F_1 = F_2^T F_2$, i.e., F_1 and F_2 give rise to the same C tensor. We have to show that

$$\hat{\Psi}(X,F_1,\Theta) = \hat{\Psi}(X,F_2,\Theta).$$

Then $\hat{\Psi}(X,C,\Theta)$ will be well defined as this common value.

Choose a regular map $\xi : N \longrightarrow N$ with $\xi(x_1) = x_2$ and with $T\xi(x_1)$ arbitrary. In particular, we can choose $T\xi(x_1)$ such that $T\xi(x_1)F_1 = F_2$ since F_1 and F_2 are assumed invertible. The assumption $F_1^T F_1 = F_2^T F_2$ implies that $T\xi(x_1)$ is an isometry. Indeed,

$$
\begin{aligned}
\langle T\psi(x_1)\cdot F_1\cdot V_1,\ T\psi(x_1)\cdot F_2\cdot V_2 \rangle &= \langle F_2\cdot V_1,\ F_2\cdot V_2 \rangle \\
&= \langle F_2^T F_2 V_1,\ V_2 \rangle \\
&= \langle F_1^T F_1 V_1,\ V_2 \rangle \\
&= \langle F_1 V_1,\ F_1 V_2 \rangle . \quad \square
\end{aligned}
$$

The last part of the proof can also be seen in coordinates as follows:

$$
\begin{aligned}
C_{AB} &= g_{ab} F^a_{1A} F^b_{1B} && (C = F_1^T F_1) \\
&= g_{ab} F^a_{2A} F^b_{2B} && (F_1^T F_1 = F_2^T F_2) \\
&= g_{ab} \frac{\partial \xi^a}{\partial x^c} F^c_{1A} \frac{\partial \xi^b}{\partial x^d} F^d_{1B} && (T\Psi \cdot F_1 = F_2)
\end{aligned}
$$

Thus, comparing the first and third lines, $g_{ab} = g_{cd} \frac{\partial \xi^c}{\partial x^a} \frac{\partial \xi^d}{\partial x^b}$ (evaluated at x_1) so ξ is an isometry, i.e., leaves the metric tensor g_{ab} invariant.

If we write the axiom of entropy production in terms of P and C by writing

$$
T : \frac{\partial F}{\partial t} = P : D = \tfrac{1}{2} P : \frac{\partial C}{\partial t}
$$

we find, instead of $\hat{T} = \rho_0 \frac{\partial \hat{\Psi}}{\partial F} \cdot g^\sharp$, the important identity

$$
\hat{P} = 2\rho_0 \frac{\partial \hat{\Psi}}{\partial C} .
$$

Remark   One can allow the constitutive functions to depend on derivatives of higher order than F and still be consistent with the entropy production inequality provided one postulates the existence of higher order stresses. This is the 'multipolar' or 'higher order' theory. (See Green and Rivlin [102].)

If we introduce the metric g on N as a variable in ψ, then the equation

$$\hat{P} = 2\rho_0 \frac{\partial \hat{\Psi}}{\partial C}$$

is equivalent to

$$\underset{\sim}{t} = 2\rho \frac{\partial \Psi}{\partial g},$$

since $\psi(c,g,\theta) = \Psi(G,C,\Theta)$. Indeed, by the chain rule,

$$\begin{aligned}
\frac{\partial}{\partial g_{ab}} \Psi(G,C,\Theta) &= \frac{\partial}{\partial g_{ab}} \Psi(G,\phi^*g,\Theta) \\
&= \frac{\partial \Psi}{\partial C_{AB}} \frac{\partial}{\partial g} (\phi^* g)_{AB} \\
&= \frac{\partial \Psi}{\partial C_{AB}} F^a{}_A F^b{}_B \\
&= \phi_* \frac{\partial \Psi}{\partial C}.
\end{aligned}$$

Therefore,

$$\begin{aligned}
\underset{\sim}{t} &= \frac{1}{J} \phi_* P \\
&= \frac{1}{J} \phi_* \frac{\partial \Psi}{\partial C} \\
&= 2\rho \frac{\partial \psi}{\partial g}.
\end{aligned}$$

Thus we see that our formula for the Cauchy stress, originally derived from a covariance assumption on the energy balance is also a deduction from constitutive theory. (The original derivation did not make any constitutive assumptions other than the dependence on the metric g on our space, which has nothing to do with any constitutive assumptions on the motion.)

Before proceeding to the covariant version of material frame indifference, we recall why $\partial\psi/\partial g$ (with variables $g, \theta \ldots\ldots c$ suppressed) is the same as $\partial e/\partial g$ (with variables g, η), assuming $\eta = -(\partial\psi/\partial\theta)$. Namely, the relationship between ψ and e is exactly the Legendre transform (up to signs):

$$e = \psi + \theta\eta.$$

On differentiating with respect to g with η held fixed:

$$\frac{\partial e}{\partial g} = \left(\frac{\partial\psi}{\partial g} + \frac{\partial\psi}{\partial\theta}\frac{\partial\theta}{\partial g}\right) + \frac{\partial\theta}{\partial g}\eta,$$

and the last two terms cancel. There is a lot of geometry behind the Legendre transform, as we know from classical mechanics (see, e.g. Abraham and Marsden [1]). This geometry, naturally, can be used in thermodynamics as well (see, e.g., Hermann [122] and Souriau [233]).

Let us start fresh and work spatially. For simplicity, we work with elasticity, the generalization to thermoelasticity being obvious. Let us make the constitutive assumption that, besides depending on the metric g, as discussed in Section 2, the internal energy e also depends on the motion ϕ and does so locally. From covariance of energy balance we have already seen that e can depend only on the point values of g and that $\underset{\sim}{t} = 2\rho(\partial e/\partial g)$. Let $\hat{E}(\phi,g) = \hat{e}(\phi,g)\circ\phi^{-1}$ (we suppress G in $\hat{E}$).

As noted above, from the first law we can already conclude that $\hat{E}$ depends only on ϕ through F.

<u>Covariant Axiom of Material Frame Indifference</u> For any regular map $\xi : N \longrightarrow N$ (i.e., a superposed configuration), we have

$$\hat{E}(\xi\circ\phi, g) = \hat{E}(\phi,\xi^*g).$$

<u>Theorem</u> The covariant axiom of MFI holds if and only if E depends on ϕ only through the point values of C.

The idea here is that the presence of the argument g and our freedom to change the metric means we are not confined to rigid motions (isometries).

<u>Proof</u>. Suppose $\hat{E}(\phi,g) = \hat{E}(C)$ (usual abuse of notation). Then

$$\begin{aligned}\hat{E}(\xi\circ\phi, g) &= \hat{E}((\xi\circ\phi)^*g)\\ &= \hat{E}(\phi^*\cdot\xi^*g)\\ &= \hat{E}(\phi,\xi^*g)\end{aligned}$$

so the covariant MFI axiom is necessary. It is sufficient, for suppose $\phi_1^*g = \phi_2^*g$ at X. Then letting $\xi = \phi_1\circ\phi_2^{-1}$, we see that $\xi^*g = g$ at $x = \phi_2(X)$. Hence,

$$\begin{aligned}\hat{E}(\phi_1,g)(X) &= \hat{E}(\xi\circ\phi_2, g)(X)\\ &= \hat{E}(\phi_2,\xi^*g)(X)\\ &= \hat{E}(\phi_2,g)(X),\end{aligned}$$

and so $\hat{E}$ depends only on $\phi^*g(X)$. □

As we have seen, a fully covariant constitutive theory can be developed for elasticity without reference to thermodynamics. For thermo-elasticity, a full constitutive theory, including the thermodynamic relation $\eta = -(\partial\psi/\partial\theta)$ depends only on ideas of covariance of energy balance, covariant MFI and on covariance of entropy production. The Clausius-Duhem inequality is not needed.

Most workers in continuum mechanics seem to agree that the foundations of the subject can be looked at in many different ways. We illustrate this by looking at the covariant form of MFI from another viewpoint suggested by M. Gurtin. This will also serve to shed light on what we have already done. Here we keep the metric g on N <u>fixed</u> and revert to isometries but we look at g in all possible coordinate charts.

We let $X_0 \in M$. A <u>local deformation</u> is a local diffeomorphism ϕ of a neighborhood N of X onto a neighborhood of X_0 such that $\phi(X_0) = X_0$. A <u>coordinate map</u> χ is a local diffeomorphism of a neighborhood of X_0 onto a neighborhood of $0 \in \mathbb{R}^n$ with $\chi(X_0) = 0$. Let

Φ = the set of local deformations,

$\mathfrak{X}$ = the set of coordinate maps.

$\Phi_{\mathfrak{X}} = \{\chi \circ \phi \mid \phi \in \Phi\}$ (independent of the choice of $\chi \in \mathfrak{X}$).

We assume given a function $E : \Phi \longrightarrow \mathbb{R}$; $E(\phi)$ is the <u>internal energy</u> at X_0 when the material around X_0 is subjected to ϕ. Choose $\chi \in \mathfrak{X}$ and consider the map $\mathcal{E}(\cdot,\chi) : \Phi_{\mathfrak{X}} \longrightarrow \mathbb{R}$ defined by $\mathcal{E}(\lambda,\chi) = E(\chi^{-1} \circ \lambda)$, i.e.,

$$\mathcal{E}(\chi \circ \phi, \chi) = E(\phi). \tag{1}$$

$\mathcal{E}(\cdot,\chi)$ is the internal energy relative to χ. A trivial consequence of (1) is the <u>coordinate covariance property</u>

$$\mathcal{E}(\chi\circ\phi,\chi) = \mathcal{E}(\mu\circ\phi,\mu)$$

for $\chi,\mu \in \mathfrak{X}$, $\phi \in \Phi$. Equivalently,

$$\mathcal{E}(\chi\circ\psi\circ\phi,\chi\circ\phi) = \mathcal{E}(\chi\circ\phi,\chi).$$

As above, we can define E to be frame-indifferent if and only if, $E(\phi) = E(\omega\circ\phi)$ for all $\phi \in \Phi$ and ω a spatial isometry at X_0. The following is readily verified.

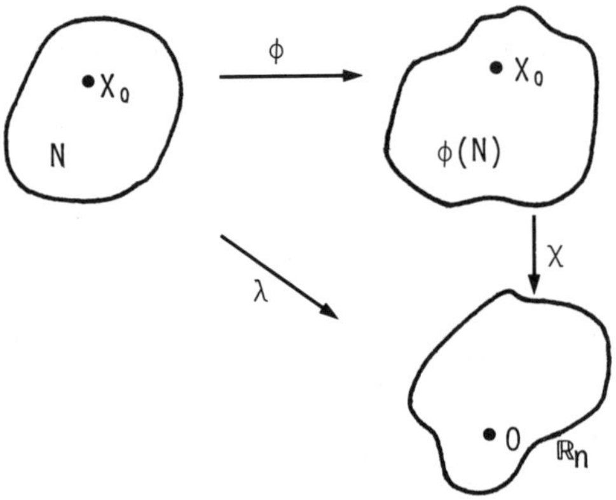

Fig. 5.

Proposition The following are equivalent:

(i) E is frame-indifferent,

(ii) $\mathcal{E}(\chi\circ\omega\circ\phi,\chi) = E(\chi\circ\phi,\chi)$ for all $\chi \in \mathfrak{X}$, $\phi \in \Phi$, ω an isometry at X_0,

(iii) $\mathcal{E}(\cdot,\chi) = \mathcal{E}(\cdot,\chi\circ\omega)$ for all $\chi \in \mathfrak{X}$, ω an isometry at X_0.

Let us call E simple if $E(\phi) = E(\alpha)$ whenever $D\phi(X_0) = D\alpha(X_0)$. Of course, we saw above that this can be deduced from thermodynamic assumptions. Arguing as we did in the theorem following the second axiom of constitutive theory, we may prove:

Proposition The following are equivalent:

(i) E is simple,

(ii) $E(\lambda,\cdot) = E(\kappa,\cdot)$ whenever $D\lambda(X_0) = D\kappa(X_0)$,

(iii) $E(\cdot,\chi) = E(\cdot,\mu)$ whenever $D\chi(X_0) = D\mu(X_0)$.

Then we can conclude, by arguments like those given before:

Theorem E is simple and frame indifferent if and only if E has the form

$$E(\lambda,\chi) = \hat{E}(C_\chi, g_\chi)$$

where $C = \phi^*g$ and g_χ, C_χ are the representations of the fixed metric g and the tensor C in the coordinate chart χ.

From the present local representation point of view then, we can write

$$E(\phi) = \hat{E}(\phi_\chi^* g_\chi, g_\chi)$$

where $\phi_\chi = \chi\circ\phi\circ\chi^{-1}$ is the local representative of χ.

The fact that $E(\phi)$ is well-defined, independent of how we represent the metric g or the deformation ϕ in coordinates χ reflects the coordinate covariance. This formulation again suggests that one ought to admit all possible metrics as genuine variables. (It is tempting to differentiate this relation with respect to g_χ in order to relate $\partial\hat{E}/\partial C$ to $\partial\hat{E}/\partial g$, but we cannot do so until we allow the metric itself and not merely its local representatives to vary.)

It is useful to compare and contrast, from the covariant point of view, the notions of material symmetry, spatial symmetry and material frame indifference for an elastic body.

A material symmetry at $X_0 \in M$ is a regular mapping $\lambda : M \longrightarrow M$, $\lambda(X_0) = X_0$

such that

$$\hat{\Psi}(X_0,\lambda^*C,\lambda^*G)J(\lambda) = \hat{\Psi}(X_0,C,G)$$

as a function of its arguments, where $J(\lambda)$ is the Jacobian of λ. (If λ is an isometry of G, this reduces to the standard definition.)

A <u>spatial symmetry</u> at $x_0 \in N$ is a regular mapping $\xi : N \longrightarrow N$, $\xi(x_0) = x_0$ such that

$$\hat{\Psi}(x_0,\xi^*C,\xi^*g)J(\xi) = \hat{\Psi}(x_0,c,g).$$

These are of a different character than material frame indifference. The latter dealt with how the material stored energy behaves under observer changes, while these symmetries reflect genuine symmetries in the material and lead, as we shall see in Section 10, to conservation laws in the form of differential identities.

For completeness, we now recall the standard representation theorems for $\hat{\Psi}$ and $\hat{P}$ for isotropic materials.

<u>Definition</u> A constitutive function $\hat{\Psi}$ on an open set $M \subset \mathbb{R}^3$ is called <u>isotropic</u> if for every $X_0 \in M$, and element $A \in O(3)$ (the orthogonal group on $\mathbb{R}^3$) there is a material symmetry λ at X_0 such that $D\lambda(X_0) = A$.

Recall that the <u>invariants</u> of an invertible symmetric matrix C are defined by

$$I_1(C) = \operatorname{tr} C, \quad I_2(C) = \det C \operatorname{tr} C^{-1} \quad \text{and} \quad I_3(C) = \det C.$$

<u>Proposition</u> The invariants of C are related to the coefficients in the characteristic polynomial $P(\lambda)$ of C as follows:

$$P(\lambda) = \lambda^3 - I_1(C)\lambda^2 + I_2(C)\lambda - I_3(C).$$

In terms of the eigenvalues $\lambda_1, \lambda_2, \lambda_3$, we have

$$I_1(C) = \lambda_1 + \lambda_2 + \lambda_3,$$

$$I_2(C) = \lambda_1\lambda_2 + \lambda_1\lambda_3 + \lambda_2\lambda_3,$$

$$I_3(C) = \lambda_1\lambda_2\lambda_3,$$

(the elementary symmetric functions of $\lambda_1, \lambda_2, \lambda_3$). Moreover, the following formula for I_2 holds:

$$I_2(C) = \tfrac{1}{2}[\operatorname{tr}(C)^2 - \operatorname{tr}(C^2)].$$

This is trivial to verify by using an orthonormal basis in which C is diagonal, noting that I_1, I_2, I_3 are rotationally invariant.

Since $\lambda_1, \lambda_2, \lambda_3$ completely determine the characteristic polynomial and hence the invariants and vice versa, these remarks also yield:

<u>Proposition</u> A scalar function f of C is invariant under orthogonal transformations if and only if f is a function of the invariants of C.

Thus for isotropic thermoelastic materials we can regard $\hat{\Psi}$ as a function of X, I_1, I_2, I_3, and Θ. Note that the number of arguments of $\hat{\Psi}$ in the C variable is thus reduced from 6 to 3.

We next compute the second Piola-Kirchhoff stress tensor in terms of this data.

<u>Theorem</u> For isotropic thermo-elastic materials, the following constitutive relation holds:

$$\hat{P}^{\flat} = \alpha_0 G + \alpha_1 C + \alpha_2 C^2, \qquad ({}^{\flat} = \text{with indices down})$$

where α_i, $i = 0,1,2$ are scalar functions of X, the invariants of C and Θ, and where $(C^2)_{AB} = C_A{}^D C_{DB}$.

Proof. We have $\hat{P} = 2\rho_0(\partial\hat{\Psi}/\partial C)$. Now,

$$\frac{\partial\hat{\Psi}}{\partial C} = \frac{\partial\hat{\Psi}}{\partial I_1}\frac{\partial I_1}{\partial C} + \frac{\partial\hat{\Psi}}{\partial I_2}\frac{\partial I_2}{\partial C} + \frac{\partial\hat{\Psi}}{\partial I_3}\frac{\partial I_3}{\partial C}.$$

But $I_1(C) = C_{AB}G^{AB}$ so $\partial I_1/\partial C = G^{\sharp}$, i.e., $\partial I_1/\partial C_{AB} = G^{AB}$.
Next we use:

Lemma $\partial I_3/\partial C = (\det C)\cdot C^{-1} = I_3(C)C^{-1}$.

Proof. From the definition of determinant, for fixed A we have

$$\det C = \varepsilon^{BCD}C_{A_1B}C_{A_2C}C_{A_3D}$$

where $\varepsilon^{BCD} = \pm 1$ depending on whether (B,C,D) is an even or odd permutation of $(1,2,3)$ and where (A_1,A_2,A_3) is a fixed even permutation of $(1,2,3)$. Thus,

$$\frac{\partial}{\partial C_{AB}}(\det C) = \varepsilon^{BCD}C_{A_2C}C_{A_3D}$$

$$= \varepsilon^{BCD}C_{AE}C_{A_2C}C_{A_3D}(C^{-1})^{AE}$$

$$= (\det C)\delta^B{}_E C^{AE}$$

$$= (\det C)(C^{-1})^{AB}. \quad \square$$

Now we can compute

$$\frac{\partial I_2}{\partial C} = \left(\frac{\partial}{\partial C}\det C\right)\operatorname{tr} C^{-1} + \det C\frac{\partial}{\partial C}\operatorname{tr} C^{-1}$$

$$= (\det C)C^{-1}\operatorname{tr} C^{-1} + (\det C)\operatorname{tr}\left(\frac{\partial C^{-1}}{\partial C}\right)$$

$$= I_2(C)C^{-1} + I_3(C)\operatorname{tr}\left(\frac{\partial C^{-1}}{\partial C}\right).$$

$\underline{\text{Lemma}}$ $\quad (\partial C^{-1}/\partial C)\cdot H = -C^{-1}\cdot H\cdot C^{-1}$, i.e., $\partial(C^{-1})^{AB}/\partial C_{CD} = -(C^{-1})^{AC}(C^{-1})^{DB}$.

Proof. Differentiate the identity $C\cdot C^{-1} = \mathrm{Id}$ in the direction H to get

$$H\cdot C^{-1} + C\cdot\frac{\partial C^{-1}}{\partial C}\cdot H = 0,$$

which gives the result. □

We now note that

$$\mathrm{tr}(\partial C^{-1}/\partial C)^{CD} = (C^{-1})^{AC}(C^{-1})^{D}{}_{A}$$

or

$$\mathrm{tr}(\partial C^{-1}/\partial C) = -C^{-2},$$

and so

$$\partial I_2/\partial C = I_2(C)C^{-1} - I_3(C)C^{-2}.$$

Substitution of these formulae yields:

$$\hat{P} = 2\rho_0\left(\frac{\partial\hat{\Psi}}{\partial I_1}G^\sharp + \left(\frac{\partial\hat{\Psi}}{\partial I_2}I_2 + \frac{\partial\hat{\Psi}}{\partial I_3}I_3\right)C^{-1} - \frac{\partial\hat{\Psi}}{\partial I_2}I_3\,C^{-2}\right).$$

From the Cayley-Hamilton theorem from linear algebra, C satisfies its characteristic equation:

$$C^3 - I_1(C)C^2 + I_2(C)C - I_3(C) = 0.$$

Thus, $C^{-1} = \dfrac{1}{I_3(C)}\{C^2 - I_1(C)C + I_2(C)\}$,

and $\quad C^{-2} = \dfrac{1}{I_3(C)}\{C - I_1(C) + I_2(C)C^{-1}\}$

$$= \frac{}{I_3(C)}\left\{\frac{I_2(C)}{I_3(C)}C^2 + \left[\mathrm{Id} - \frac{I_1(C)I_2(C)}{I_3(C)}\right]C + \left[\frac{I_2(C)}{I_3(C)} - I_1(C)\right]\right\}.$$

Inserting these expressions into the above formula for $\hat{P}$ yields the desired conclusion.

Next we turn our attention to the elasticity tensors; we shall use these notions in our discussions of linearization and for the existence and uniqueness theory.

From balance of momentum we have from Section 2

$$\rho_0 A = \rho_0 B + \mathrm{DIV}\,T, \quad \text{i.e.,} \quad \rho_0\left(\frac{\partial V^a}{\partial t} + \gamma^a_{bc}V^b V^c\right) = \rho_0 B^a + T^{Aa}{}_{|A}.$$

If we use the constitutive hypothesis for a thermoelastic material, we can write T as a function $\hat{T}$ of X, F and Θ. Then we can compute $\mathrm{DIV}\,T$ by the chain rule, as follows:†

$$\mathrm{DIV}\,T = \frac{\partial \hat{T}}{\partial F}\cdot \nabla_X F + \mathrm{DIV}_X \hat{T} + \frac{\partial \hat{T}}{\partial \Theta}\cdot \nabla_X \Theta,$$

where $\mathrm{DIV}_X T$ means the divergence of T holding the variables F and Θ constant. In coordinates,

$$(\mathrm{DIV}\,T)^a = \frac{\partial \hat{T}^{Aa}}{\partial F^b{}_B} F^b{}_{B|A} + \left(\frac{\partial T^{Aa}}{\partial X^A} + \hat{T}^{Aa}\Gamma^B{}_{AB} + \hat{T}^{Ab}\gamma^a{}_{bc}F^c{}_A\right) + \frac{\partial \hat{T}^{Aa}}{\partial \Theta}\frac{\partial \Theta}{\partial X^A},$$

where we have written out $\mathrm{DIV}_X T$ explicitly using the formula for the

---

†Actually there are subtleties involved here. $\hat{T}$ is a vector bundle mapping $\hat{T} : \mathcal{E} \longrightarrow \mathcal{F}$ where $\mathcal{E}$ is the bundle over M whose fiber at X consists of linear maps of $T_X M$ to $T_{\phi(X)}N$ direct sum the scalars $(\mathbb{R})$, and $\mathcal{F}$ is the bundle of two point tensors over ϕ. The fiber derivative $\partial\hat{T}/\partial F$ makes perfectly good sense, but to write $\mathrm{DIV}_X \hat{T} = \mathrm{tr}(\nabla_X \hat{T})$, one has to put a connection on $\mathcal{E}$ and $\mathcal{F}$, compute $\nabla\hat{T}$ and take its horizontal part. This process is actually equivalent to the version given in the text.

covariant derivative of a two point tensor*, and where

$$F^b{}_{B|A} = \frac{\partial^2\phi^b}{\partial X^A \partial X^B} + \frac{\partial\phi^e}{\partial X^B}\gamma^b{}_{ec}\frac{\partial\phi^c}{\partial X^A} - \frac{\partial\phi^b}{\partial X^C}\Gamma^C{}_{AB}\,.$$

(Recall that γ^a_{bc} is evaluated at $x = \phi(x)$). Thus the leading term in DIV T, containing second derivatives of ϕ is

$$\frac{\partial\hat{T}^{Aa}}{\partial F^b{}_B}\frac{\partial^2\phi^b}{\partial X^A\partial X^B} = \frac{1}{2}\left(\frac{\partial\hat{T}^{Aa}}{\partial F^b{}_B} + \frac{\partial\hat{T}^{Ba}}{\partial F^b{}_A}\right)\frac{\partial^2\phi^b}{\partial X^A\partial X^B}\,.$$

This leads to the following

<u>Definition</u> Let $\hat{T}$ be a constitutive function for thermo-elasticity, depending on X, F and Θ. Define the (<u>first</u>) <u>elasticity tensor</u> A by

$$A = \frac{\partial\hat{T}}{\partial F}\,,$$

i.e.,

$$A^{ABa}{}_b = \frac{\partial\hat{T}^{Aa}}{\partial F^b{}_B}$$

so that A is a two-point tensor $\left[\text{of type } \begin{pmatrix}2 & 1\\ 0, & 1\end{pmatrix}\right]$.

We shall also write $A^\flat$ for A with its third index lowered, i.e., for $A^{AB}{}_{ab}$ and shall write A_S for A symmetrized on its large indices, i.e.,

$$A^{(AB)a}{}_b = \frac{1}{2}\left(A^{ABa}{}_b + A^{BAa}{}_b\right);$$

$A^\flat{}_S$ is defined similarly.

---

*We do not know a mathematics reference which gives a discussion of two point tensors from the modern point of view. However it is a straightforward exercise to do so using the same theory as for ordinary tensors (see Bishop and Goldberg [26]). Details are given in Hughes and Marsden [132]. The correct coordinate formulae and basic ideas from which to start may be found in Ericksen's appendix to Truesdell and Toupin [249].

In order to exploit the symmetries provided by material frame indifference and balance of moment of momentum it is also convenient to work with the second Piola-Kirchhoff stress tensor. This leads to the following:

<u>Definition</u> Let $\hat{P}$ be a constitutive function depending on X, C and Θ, as above. Then the tensor on M defined by

$$\mathbb{C} = \frac{\partial \hat{P}}{\partial C}$$

is called the (<u>second</u>) <u>elasticity</u> <u>tensor</u>, i.e.,

$$C^{ABCD} = \frac{\partial \hat{P}^{AB}}{\partial C_{CD}} .$$

Notice that $\mathbb{C}$ is a fourth order tensor on M, i.e., it is not a two point tensor and does not depend on the configuration.

From the axiom of entropy production (which we assume), we saw that $\hat{P}^{AB} = 2\rho_0(\partial\hat{\Psi}/\partial C_{AB})$, so we get an important formula:

<u>Proposition</u> $C^{ABCD} = 2\rho_0(\partial^2\hat{\Psi}/\partial C_{AB}\partial C_{CD})$ and so we have the symmetries:

$$C^{ABCD} = C^{BACD} = C^{ABDC} = C^{CDAB}.$$

We can relate $\mathbb{D}$ and $\mathbb{C}$ using the formula $T = PF^T$, i.e., $T^{Aa} = P^{AB}F^a{}_B$.

<u>Proposition</u> The following formulae hold in general coordinates

(a) $A^{ABa}{}_b = 2C^{ACBD}F^c{}_D F^a{}_C g_{cb} + \hat{P}^{AB}\delta^a{}_b$,

(b) $A^{AB}{}_{ab} = 2C^{ACBD}F^c{}_D F^d{}_C g_{cb} g_{da} + \hat{P}^{AB} g_{ab}$,

(c) $A^{AB}{}_{ab} = A^{BA}{}_{ba}$.

Proof. We have

$$\frac{\partial \hat{T}^{Aa}}{\partial F^b{}_B} = \frac{\partial \hat{P}^{AC}}{\partial C_{DE}} \frac{\partial C_{DE}}{\partial F^b{}_B} F^a{}_C + \hat{P}^{AC} \frac{\partial F^a{}_C}{\partial F^b{}_B} .$$

From $C_{DE} = F^d{}_D F^c{}_E g_{dc}$ we get

$$\frac{\partial C_{DE}}{\partial F^b{}_B} = \delta^d{}_b \delta^B{}_D F^c{}_E g_{dc} + F^d{}_D \delta^c{}_b \delta^B{}_E g_{dc}$$

$$= \delta^B{}_D F^c{}_E g_{bc} + F^d{}_D \delta^B{}_E g_{db}.$$

By substitution, we obtain

$$\frac{\partial \hat{T}^{Aa}}{\partial F^b{}_B} = C^{ACDE}\left(\delta^B{}_D F^c{}_E g_{bc} + F^d{}_D \delta^B{}_E g_{db}\right) + \hat{P}^{AC} \delta^a{}_b \delta^B{}_C$$

$$= C^{ACBE} F^c{}_E g_{bc} + C^{ACDB} F^d{}_D g_{db} + \hat{P}^{AB} \delta^a{}_b.$$

On using the symmetry $C^{ACBE} = C^{ACEB}$, (a) follows. Part (b) follows by lowering the third index and (c) follows from (b) using the symmetry $C^{ACBD} = C^{BDAC}$. □

Notice especially that the tensor $A^{\flat}$ is not necessarily symmetric in each pair of indices AB and ab separately, but only when both pairs are simultaneously transposed. In three dimensions it is easy to see that the tensors with this symmetry form (pointwise) a space of dimension 54. However the dimension of the tensors with the symmetries of the $\mathbb{C}$ tensor is only 21. Thus for the second elasticity tensor there is less to keep track of, in principle.

Often P or T are taken as primitive objects and are not necessarily assumed to be derived from a free energy function Ψ. Notice that C^{ABCD} is always symmetric in AB and CD separately.

From Vainbergs' theorem proved in Section 1, (together with the observation that the set of variables C_{AB} is an open convex cone) we see that the symmetry condition

$$C^{ABCD} = C^{CDAB}, \quad \text{i.e.,} \quad A^{AB}{}_{ab} = A^{BA}{}_{ba}$$

is <u>equivalent</u> to the existence of a free energy function. When this condition is dropped, the theory is known as <u>Cauchy elasticity</u>. There are physical arguments (not always agreed upon) based on thermodynamics which say that Cauchy elasticity may be physically unreasonable. In Section 7 we shall prove under some technical conditions that in the <u>linear</u> theory the equations of Cauchy elasticity cannot be well-posed (in the semigroup sense); i.e., <u>well-posedness implies the aforementioned symmetry</u>.

When the free energy (or these symmetries) is assumed to exist, the phrase <u>hyperelasticity</u> is sometimes used in distinction to Cauchy elasticity.

When thermal effects are ignored, i.e., when Θ is omitted, we are in the case of isothermal hyper-elasticity. We shall mean this if we just say 'elasticity' in future. In this case the free energy $\hat{\Psi}$ coincides with the internal energy $\hat{E}$ and is sometimes denoted W. Thus W will be a function of (X,C) and we will have

$$\hat{T} = \rho_0 \frac{\partial W}{\partial F} \cdot g, \qquad \hat{P} = 2\rho_0 \frac{\partial \overline{W}}{\partial C}, \qquad \text{etc.}$$

We sometimes speak of W as the <u>stored energy function</u>.

In the nonlinear theory the form possible for the elasticity tensor can be discouragingly complex. For example for isotropic materials the elasticity tensor $\mathbb{C}$ has the following component form (after a healthy computation):

$$C^{ABCD} = \gamma_1 \cdot G^{AB}G^{CD} + \gamma_2 \cdot \{C^{AB}G^{CD} + G^{AB}C^{CD}\} + \gamma_3\{(C^2)^{AB}G^{CD} + G^{AB}(C^2)^{CD}\}$$

$$+ \gamma_4 \cdot C^{AB}C^{CD} + \gamma_5 \cdot \{(C^2)^{AB}C^{CD} + C^{AB}(C^2)^{CD}\} + \gamma_6 \cdot (C^2)^{AB}(C^2)^{CD}$$

$$+ \gamma_7\{G^{AC}G^{BD} + G^{BC}G^{AD}\} + \gamma_8\{G^{AC}C^{BD} + G^{BC}C^{AD} + G^{AD}C^{BC} + G^{BD}C^{AC}\},$$

where $\gamma_1, \ldots, \gamma_8$ are scalar functions of X, the invariants of C and, if the material is thermoelastic, Θ. For this reason most work on specific problems relies on drastically simplified models (such as a Mooney-Rivlin material[†]) or the linearized theory described in Section 4.

Now we shall return to the equations of motion for a thermo-elastic material, insert the first elasticity tensor, and formulate a couple of basic boundary value problems. A considerable portion of our work in the ensuing sections will be devoted to these problems.

The following notation will be convenient. The vector $B_I = \mathrm{DIV}_X \hat{T}$, i.e.,

$$B_I{}^a = \frac{\partial \hat{T}^{Aa}}{\partial X^A} + \hat{T}^{Aa}\Gamma^B{}_{AB} + \hat{T}^{Ab}\gamma^a{}_{bc}F^c{}_A,$$

is called the <u>resultant force due to inhomogeneities</u>. (B_I is a function of X, F and Θ).

The equation of motion now reads

$$\rho_0 \dot{V} = \rho_0 B + \mathrm{DIV}\, T$$

$$= \rho_0 B + A \cdot \nabla_X F + B_I + \frac{\partial \hat{T}}{\partial \Theta}\frac{\partial \Theta}{\partial X}.$$

This equation will be thought of as governing the evolution of the configuration ϕ and is coupled to the equation of energy balance which in

---

[†]See, e.g., Truesdell and Noll [248], pp. 349-355.

turn governs the evolution of Θ.

In addition to these evolution equations, some boundary conditions must be imposed. For each of ϕ and Θ there are at least three types in common use:

Definition

(I) Boundary Conditions for ϕ

(a) displacement: ϕ is prescribed (equals a given displacement) on ∂M, the boundary of M,

(b) traction: the tractions $\tau^a = \langle N, \hat{T}\rangle^a = N_A T^{Aa}$ are prescribed on ∂M,

or (c) mixed: ϕ is prescribed on a part ∂_1 of ∂M and $\langle N, \hat{T}\rangle$ on part ∂_2 of ∂M where $\partial_1 \cap \partial_2 = \emptyset$ and $\overline{\partial_1 \cup \partial_2} = \partial M$.

(II) Boundary Conditions for Θ

(a) prescribed temperature: Θ is prescribed on ∂M (Dirichlet boundary conditions),

(b) prescribed flux: $\langle N, \hat{Q}\rangle$ is prescribed on ∂M (Neuman boundary conditions),

or (c) mixed: Θ is prescribed on a part ∂_3 of ∂M and $\langle N, \hat{Q}\rangle$ on another part ∂_4 of ∂M where $\partial_3 \cap \partial_4 = \emptyset$, $\overline{\partial_3 \cup \partial_4} = \partial M$.

Notice that conditions I(b) and II(b) are, in general, nonlinear boundary conditions because $\hat{T}$ and Q are nonlinear functions of F and Θ.

Definition By the initial boundary value problem for thermo-elasticity, we mean the problem of finding $\phi(t,X)$ and $\Theta(t,X)$ such that

(i) $\rho_0 \dot{V} = A \cdot \nabla_X F + \rho_0 B + B_I + \dfrac{\partial \hat{T}}{\partial \Theta} \dfrac{\partial \Theta}{\partial X}$,

(ii) $\rho_0 \Theta \frac{\partial \hat{N}}{\partial t} + \mathrm{DIV}\, \hat{Q} = \rho_0 R,$

(iii) boundary conditions (I) and (II) hold and

(iv) ϕ, V and Θ are given at $t = 0$ (<u>initial conditions</u>) where $\hat{\Psi}$ is a given constitutive function depending on X, C and Θ and $\hat{N}$, $\hat{T}$, A, B_I are given in terms of it as above, where B, ρ_0, R are given and $\hat{Q}$ is a given function of X, F, Θ and $\mathrm{GRAD}\,\Theta$ satisfying $\langle \hat{Q}, \mathrm{GRAD}\,\Theta \rangle \leqslant 0$.

<u>Remark</u> If the motions are not sufficiently differentiable and shocks can develop, this has to be supplemented by the entropy production inequality (see, e.g., Lax [169]).

<u>Definition</u> If we omit Θ, the corresponding equation (ii), the term $\frac{\partial \hat{T}}{\partial \Theta} \frac{\partial \Theta}{\partial X}$ in (i) and the boundary and initial conditions for Θ, the resulting problem for determination of ϕ is called the <u>initial boundary value problem for elasticity</u> (or hyper-elasticity).

Notice that in (ii),

$$\begin{aligned}
\rho_0 \frac{\partial \hat{N}}{\partial t} &= \rho_0 \frac{\partial \hat{N}}{\partial C} \frac{\partial C}{\partial t} + \rho_0 \frac{\partial \hat{N}}{\partial \Theta} \frac{\partial \Theta}{\partial t} \\
&= -\rho_0 \frac{\partial}{\partial C} \frac{\partial \hat{\Psi}}{\partial \Theta} \cdot 2D + \rho_0 \frac{\partial^2 \hat{\Psi}}{\partial \Theta^2} \frac{\partial \Theta}{\partial t} \\
&= - \frac{\partial \hat{P}}{\partial \Theta} \cdot D + \rho_0 \frac{\partial^2 \hat{\Psi}}{\partial \Theta^2} \frac{\partial \Theta}{\partial t},
\end{aligned}$$

where D is the rate of deformation tensor.

<u>Definition</u> The <u>boundary value problem for thermo-elastostatics</u> consists of finding ϕ and Θ as functions of X alone such that

(i) $\mathrm{DIV}\, \hat{T} + \rho_0 B = A \cdot \nabla_X F + \rho_0 B + B_I + \frac{\partial \hat{T}}{\partial \Theta} \frac{\partial \Theta}{\partial X} = 0,$

(ii) $\mathrm{DIV}\, \hat{Q} = \rho_0 R,$

and (iii) boundary conditions (I) and (II) hold.

The boundary value problem for elastostatics consists of finding a (regular) deformation ϕ such that (i) and boundary conditions (I) hold.

The static problem is, of course, obtained from the dynamic one by dropping time derivatives.

Proposition The prescribed tractions $\tau = \langle N,\hat{T}\rangle$ in I(b) on ∂M must satisfy the necessary condition (using Euclidean coordinates in $\mathbb{R}^n$):

$$\int_{\partial M}\langle N,\hat{T}\rangle\, dA + \int_M \rho_0 B dV = 0,$$

if the traction boundary value problem for (thermo-) elastostatics has a (regular) solution. Similarly, the prescribed fluxes in II(b) must satisfy

$$\int_{\partial M}\langle N,\hat{Q}\rangle\, dA - \int_M \rho_0 R dV = 0.$$

Proof. This follows at once from $\mathrm{DIV}\,\hat{T} + \rho_0 B = 0$ by integration over M and use of Gauss' theorem. □

We shall have a good deal more to say about conditions of this type in Section 8 and also from the variational viewpoint in Section 10.

While all of the above may also be formulated in terms of the spatial picture, in elasticity the material (Lagrangian) picture is found most useful for studying boundary value problems. On the other hand, since we shall need the spatial forms of the elasticity tensors A and $\mathbb{C}$ in our discussion of linearization, it is convenient to define them here.

Definition The spatial elasticity tensors a and $2\mathbb{c}$ are the push-forwards or Piola transforms of the tensors A and $\mathbb{C}$. In coordinates,

$$a^{abc}{}_{d} = \frac{1}{J} F^{a}{}_{A} F^{b}{}_{B} A^{ABc}{}_{d},$$

$$c^{abcd} = \frac{1}{2J} F^{a}{}_{A} F^{b}{}_{B} F^{c}{}_{C} F^{d}{}_{D} C^{ABCD}.$$

The following relations come from the corresponding material results and the Piola identity.

<u>Proposition</u> The following hold:

(a) $a^{abc}{}_{d} = t^{ab} \delta^{c}{}_{d} + c^{abcde} g_{ed},$

(b) $\mathbf{a}$ and $\mathbf{c}$ have these symmetries:

$$a^{abcd} = a^{cdab},$$

$$c^{abcd} = c^{bacd} = c^{abdc} = c^{cdab},$$

(c) if $U^{a}{}_{B}$ is a two-point tensor field over ϕ, then

$$\left(A^{ABc}{}_{d} U^{d}{}_{B}\right)_{|A} = J\left(a^{abc}{}_{d} u^{d}{}_{b}\right)_{|a},$$

where $u^{d}{}_{b} = (F^{-1})^{B}{}_{b} U^{d}{}_{B}$ is the push-forward of U.

As discussed in Sections 2 and 3, we can regard t as a function of x and c (and Θ if temperatures are involved), just as P is a function of X and C. Then the equations of motion in the spatial picture,

$$\rho a = \operatorname{div} t + \rho b,$$

become,

$$\rho a^{b} = b^{abB}{}_{d} F^{d}{}_{B|a} + \frac{\partial t^{ab}}{\partial x^{a}} + \rho b^{b},$$

where $\mathbb{b}$ is the tensor given by

$$b^{abB}{}_d = \frac{\partial t^{ab}}{\partial F^d{}_B}.$$

We caution the reader that $2c^{abcd}$ is not $\partial t^{ab}/\partial c_{ab}$ nor is t^{ab} given by $2\rho(\partial\psi/\partial c_{ab})$ (it is $2\rho(\partial\psi/\partial g_{ab})$ as in Sections 2 and 3).

We conclude this section with a remark on incompressible elasticity. Here one imposes the constraint that ϕ be volume preserving, i.e., $J = 1$. (For instance such a condition is often imposed on rubber.) This condition is perhaps best understood in terms of Hamiltonian systems with constraints, and we discuss this point of view in Section 10. For now we merely remark that this condition introduces a Lagrange multiplier into the equations in the usual way. We replace

$$t \text{ by } t - pg^\sharp, \quad \text{i.e.,} \quad t^{ab} \text{ by } t^{ab} - pg^{ab},$$

where p is an unknown function, the <u>pressure</u>, to be determined by the condition of incompressibility. In terms of the first Piola-Kirchhoff tensor T, we replace $\hat{T}$ from our constitutive theory by

$$\hat{T} - JPF^{-1}$$

where P is a function of (t,x) to be determined by $J = 1$. We emphasize that in an initial boundary value problem, p becomes unknown and depends on ϕ in a non-local way, as in fluid mechanics. (See, e.g., Hughes and Marsden [133] and references therein.)

4 LINEARIZATION

Linearization provides a key link between the linear and nonlinear theories of elasticity. Our first goal in this section is to linearize the nonlinear equations to obtain the equations of linear elasticity. (The linear theory can, of course, also be developed separately, as in Gurtin [106]). In doing so, we shall assume the reader is familiar with calculus in Banach spaces (see, for instance, Dieudonné [68]), although we shall review the notation.

Often, linearization of the equations of continuum mechanics is done in Euclidean coordinates and then, at the end, partial derivatives are replaced by covariant derivatives. This is unsatisfactory and dangerous. Surprisingly it is not entirely trivial to give a covariant linearization procedure. One of our first goals is to do so.

Following this we shall describe how the implicit function theorem can be used to obtain local solutions to nonlinear problems from a theorem about the linearized problem. This will be applied in Section 8 to elasticity, where theorems of Stoppelli and Van Buren will be proved.

One of the precautions to be stressed is that perfectly sound nonlinear theories can have a pathological linearization in the sense that solutions of the linearized equations need <u>not</u> be first order approximations of solutions to the nonlinear equations. As was discovered by Signorini [228, 229] in 1930, this occurs in the elastostatics traction problem in the presence of an axis of equilibrium, and will be called a linearization instability. We shall briefly discuss this problem in Section 11. A main result, due to Stoppelli, describing the set of solutions near equilibrium, can be obtained and extended using ideas of generic bifurcation theory (see Chow, Hale and Mallet-Paret [47] and Hale [118]).†

---

†Another equivalent way of stating linearization instability is this: after any necessary scalings, a perturbation series may have the property that at $(n+1)^{th}$ order, corrections to n^{th} order may be necessary. If such corrections stop after finite order, we say that the singularity is <u>finitely determined</u>.

Linearization instabilities occur in nonlinear theories other than elasticity. For example, in general relativity a linearization instability occurs at a spacetime which has symmetries (which is analogous in elasticity to the case when the loads have an axis of equilibrium). See Fischer and Marsden [87].

In order to establish our notation, we recall the definition of Fréchet differentiability.

<u>Definition</u> Let X and Y be Banach spaces, $U \subset X$ be open and $f : U \subset X \longrightarrow Y$. We say f is <u>differentiable</u> at $x_0 \in U$ if there is a bounded linear operator

$$Df(x_0) \in B(X,Y),$$

(the set of all bounded linear operators from X to Y) such that for each $\varepsilon > 0$ there is a $\delta > 0$ such that $\|h\|_X < \delta$ implies

$$\|f(x_0+h) - f(x_0) - Df(x_0)\cdot h\|_Y \leqslant \varepsilon\|h\|_X .$$

(This uniquely determines $Df(x_0)$).

We say that f is C^1 if it is differentiable at each point of U and if $x \longmapsto Df(x)$ is continuous from U to $B(X,Y)$ with the <u>norm</u> topology.

In Euclidean spaces, $Df(x_0)$ is the linear map whose matrix in the standard bases is the matrix of partial derivatives of f.

The concept "f is of class C^r", $0 \leqslant r \leqslant \infty$ is defined inductively. For example, f is C^2 if it is C^1 and $x \longmapsto D^2f(x) \in B(X,L(X,Y))$, the derivative of $x \longmapsto Df(x)$, is norm continuous. The space $B(X,L(X,Y))$ is isomorphic to $B^2(X,Y)$, the space of all continuous bilinear maps $b : X \times X \longrightarrow Y$ by $b \longmapsto \hat{b}$, $\hat{b}(x_1) \cdot x_2 = b(x_1,x_2)$. Thus $D^2f(x)$ is usually regarded as a bilinear map of X to Y. Its value on

$(x_1, x_2) \in X \times X$ will be denoted $D^2f(x) \cdot (x_1, x_2)$. We recall that if f is C^2, then $D^2f(x)$ is <u>symmetric</u>. If f is a function of two (or more) variables; say $f : U \subset X_1 \times X_2 \longrightarrow Y$, the partial derivatives are denoted by D_1f and D_2f (or sometimes $D_{X_1}f$, etc.).

The rules of elementary calculus (chain rule, Taylor's theorem, etc.,) all carry over using essentially the same proofs.

In examples given below, the special case of the chain rule

$$\frac{d}{d\varepsilon} f(x_0 + \varepsilon u)\Big|_{\varepsilon=0} = Df(x_0) \cdot u,$$

relating $Df(x_0)$ to directional derivatives, is quite useful for computing Df. A map for which all the directional derivatives exist (at x_0) and comprise a bounded operator is called Gâteaux differentiable (at x_0). This slightly weaker notion of differentiability is sometimes useful.

If $f : U \subset X \longrightarrow Y$ is C^1 and we are trying to solve the equation $f(x) = 0$ then we may write $x = x_0 + h$ and approximate $f(x_0 + h)$ by the first two terms in its Taylor expansion. This leads us to the linearized equations.

<u>Definition</u> Let $f : U \subset X \longrightarrow Y$ be a C^1 map and let $x_0 \in U$ (not necessarily satisfying $f(x_0) = 0$). Then the linearization of the equations $f(x) = 0$ about x_0 are the equations

$$L(x_0, h) = f(x_0) + Df(x_0) \cdot h = 0$$

for $h \in X$.

Thus $x_0 + h$ will satisfy the nonlinear equations $f(x) = 0$ to first order. This is <u>not</u> the same as saying $x_0 + h$ is, to first order, an exact solution. For instance, let $f : \mathbb{R}^2 \longrightarrow \mathbb{R}$ be given by $f(x, y) = x^2 + y^2$, and let $x_0 = (0,0)$. Since $Df(0,0) = 0$, any (h,k) satisfies the linearized equations but is not, to first order, an exact solution of the nonlinear equations. This trivial observation is relevant for understanding the concepts of linearization stability discussed below.

Many important maps in nonlinear analysis are defined by composition. This occurs, in particular, in elasticity. The next theorem shows us how to differentiate such maps. Results of this type go back at least to Sobolev in the 1930's (see Sobolev [231]). For simplicity, we work in C^k spaces, but the same thing works in a variety of function spaces such as $W^{s,p}$ spaces. (This observation is needed later.)

Theorem Let $\bar{\Omega} \subset \mathbb{R}^n$ be a nice region and let X be the Banach space of C^k maps† $u : \bar{\Omega} \longrightarrow \mathbb{R}^m$ and let Y be the Banach space of C^{k-1} maps $g : \bar{\Omega} \longrightarrow \mathbb{R}^k$, $1 \leqslant k < \infty$. Let

$$W : \bar{\Omega} \times \mathbb{R}^m \times L(\mathbb{R}^n,\mathbb{R}^m) \longrightarrow \mathbb{R}^k$$

be C^r, $r \geqslant k$, and define

$$f : X \longrightarrow Y, \quad f(u)(x) = W(x,u(x),Du(x)).$$

Then f is of class C^r and

$$(Df(u)\cdot v)(x) = D_2W(x,u(x),Du(x))\cdot v(x) + D_3W(x,u(x),Du(x))\cdot Dv(x).$$

In components, and suppressing variables, this reads

---

†Precisely, a map $u : \bar{\Omega} = \Omega \cup \partial\Omega \longrightarrow \mathbb{R}^m$ is C^k when it has a C^k extension to an open set containing $\bar{\Omega}$. The Whitney extension theorem is relevant here.

$$(Df(u)\cdot v)^j = \frac{\partial W^j}{\partial u^i} v^i + \frac{\partial W^j}{\partial\left(\frac{\partial u^i}{\partial x^1}\right)} \cdot \frac{\partial v^i}{\partial x^1}.$$

Proof. Induction easily reduces the argument to the case $r=1$, $k=1$. The following computation and the finite dimensional chain rule shows that f is Gâteaux differentiable with derivative as stated in the theorem:

$$[\,Df(u)\cdot v\,](x) = \frac{d}{d\varepsilon} f(u+\varepsilon v)(x)\Big|_{\varepsilon=0}$$

$$= \frac{d}{d\varepsilon} W(x, u(x)+\varepsilon v(x), Du(x)+\varepsilon Dv(x))\Big|_{\varepsilon=0}.$$

A straightforward uniform continuity argument shows that $u \longmapsto Df(u) \in B(X,Y)$ is norm continuous. The proof is now completed using the following:

Lemma Let $f : U \subset X \longrightarrow Y$ be Gâteaux differentiable and assume $u \longmapsto Df(u) \in B(X,Y)$ is continuous. Then f is C^1.

Proof. By the fundamental theorem of Calculus,

$$f(u_0+h) - f(u_0) - Df(u_0)\cdot h = \int_0^1 \frac{d}{d\lambda} f(u_0+\lambda h)\,d\lambda - Df(u_0)\cdot h$$

$$= \int_0^1 [\,Df(u_0+\lambda h)\cdot h - Df(u_0)\cdot h\,]\,d\lambda.$$

By continuity of Df, for any $\varepsilon > 0$ there is a $\delta > 0$ such that $\|Df(u) - Df(u_0)\| < \varepsilon$ if $\|u - u_0\| < \delta$. Then $\|h\| < \delta$ implies:

$$\|f(u_0+h) - f(u_0) - Df(u_0)\cdot h\| \leqslant \int_0^1 \|Df(u_0+\lambda h) - Df(x_0)\| \cdot \|h\|\,d\lambda \leqslant \varepsilon\|h\|. \quad \square$$

Since f is continuous and linear in W, this shows that in fact f is

C^r as a function of the pair (u,W).

In order to obtain a covariant description of linearization in elasticity, we shall need to generalize this procedure to infinite dimensional manifolds (cf. Lang [167]).

Suppose $\mathscr{X}$ is a manifold, possibly infinite dimensional, and that $\pi : \mathscr{E} \longrightarrow \mathscr{X}$ is a vector bundle over $\mathscr{X}$. Let $f : \mathscr{X} \longrightarrow \mathscr{E}$ be a section of this bundle. We are interested in the nonlinear equation

$$f(\phi) = 0, \quad \phi \in \mathscr{X}.$$

To form the linearized equations about a given $\phi_0 \in \mathscr{X}$, we need some additional structure. Namely, we need to assume we have a connection on $\mathscr{E}$; i.e., we can parallel transport fibers of $\mathscr{E}$ along curves in $\mathscr{X}$, exactly as in Riemannian geometry. Thus we can form the covariant derivative of f, denoted ∇f. As in the finite dimensional case, the covariant derivative is defined as follows:

Let $\phi_0 \in \mathscr{X}$ and $\mathscr{E}_0 = \pi^{-1}(\phi_0)$ be the fiber over ϕ_0, a linear space. Let $V \in T_\phi \mathscr{X}$ be a tangent vector to $\mathscr{X}$ at ϕ_0 and let it be tangent to a curve $\phi_t \in \mathscr{X}$. Then we define

$$\nabla f(\phi_0)\cdot V = \frac{d}{dt}\alpha_t f(\phi_t)\Big|_{t=0},$$

where $\alpha_t : \mathscr{E}_t = \pi^{-1}(\phi_t) \longrightarrow \mathscr{E}_0$ is the parallel translation map. Thus, $\nabla f(\phi_0)$ is a linear map of $T_{\phi_0}\mathscr{X}$ to $\mathscr{E}_0$.

With this notation we are ready to define the linearized equations.

<u>Definition</u> The <u>linearization</u> of the equations $f(\phi) = 0$ at ϕ_0 are the equations

$$L(\phi_0, V) = f(\phi_0) + \nabla f(\phi_0)\cdot V = 0$$

for $V \in T_{\phi_0}\mathscr{X}$.

In charts $\nabla f(\phi_0)$ is related to $Df(\phi_0)$ via the addition of some 'Christoffel symbol' terms. It is worth observing that if $f(\phi_0) = 0$, then $\nabla f(\phi_0)$ is the same as $Df(\phi_0)$ in charts.

Now we want to apply these ideas to compute the linearized equations of elasticity. (Thermo-elasticity, etc., may be done the same way.) For simplicity we continue to work in C^k spaces although the results work in Sobolev or Hölder spaces just as well.

Let $\mathscr{C}$ be the set of all regular C^k configurations $\phi : M \longrightarrow N$. If $N = \mathbb{R}^n$, then we can, on using the Euclidean structure, regard $\mathscr{C}$ as an open set in a Banach space X. Even for general N, one can show that $\mathscr{C}$ is a C^∞ infinite dimensional manifold.† A tangent vector to $\mathscr{C}$ at $\overset{\circ}{\phi} \in \mathscr{C}$ is the tangent to a curve $\phi_t \in \mathscr{C}$ with $\phi_0 = \overset{\circ}{\phi}$, i.e., to a <u>motion</u>. Thus, a tangent vector to $\mathscr{C}$ at $\overset{\circ}{\phi}$ is a vector field U covering $\overset{\circ}{\phi}$, from our work in Section 1; see figure 6. Sometimes, U is spoken of as an

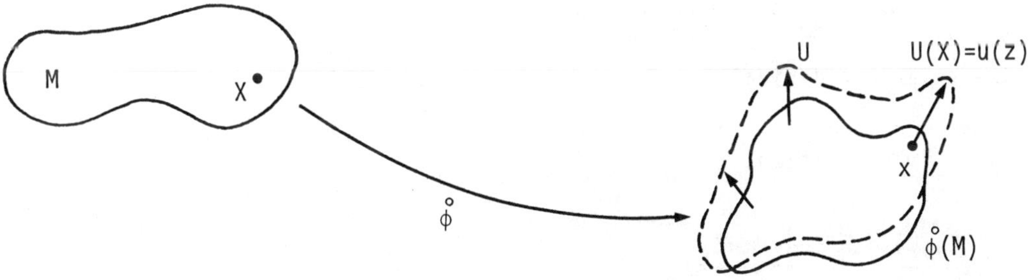

an infinitesimal displacement of a configuration $\overset{\circ}{\phi}$ is a vector field over $\overset{\circ}{\phi}$.

Fig. 6.

†See Palais [207], Ebin and Marsden [77] and references therein. Here we shall impose the boundary conditions of place or traction as separate equations. However, one may wish to include some of them in the definition of $\mathscr{C}$. Then $\mathscr{C}$ will not be open in a linear space, but will still be a manifold. This point of view is taken up in Section 10.

infinitesimal deformation imposed on the finite deformation ϕ. (See Rivlin [221].)

We shall build up the linearized equations in several steps. First of all, consider the association $\phi \mapsto F = T\phi$. Here we let $\mathcal{E}$ be the vector bundle over $\mathcal{X}$ whose fiber at $\overset{\circ}{\phi}$ consists of all C^{k-1} maps $F: TM \longrightarrow TN$ which cover $\overset{\circ}{\phi}$. There is a natural notion of parallel translation on $\mathcal{E}$ obtained by pointwise parallel translation of two point tensors F over curves in N. (In Euclidean space this operation is just ordinary translation.)

<u>Proposition</u> The linearization of the map $f: \phi \mapsto F = T\phi$ at $\overset{\circ}{\phi}$ is given by

$$L(\overset{\circ}{\phi},U) = \overset{\circ}{F} + \nabla U,$$

where $\overset{\circ}{F} = T\overset{\circ}{\phi}$. In coordinates,

$$(\nabla U)^a{}_A = U^a{}_{|A} = \gamma^a{}_{bc}U^c\overset{\circ}{F}{}^b{}_A + \frac{\partial U^a}{\partial X^A}$$

is the covariant derivative of the two point tensor U.

<u>Proof</u>. By definition,

$$L(\overset{\circ}{\phi},U) = \overset{\circ}{F} + \frac{d}{dt}\alpha_t \cdot F_t \Big|_{t=0}$$

where α_t denotes parallel translation, and $F_t = T\phi_t$, where ϕ_t is tangent to U at $t=0$; $\phi_0 = \overset{\circ}{\phi}$. At $X \in M$, $\alpha_t \cdot F_t(X)$ is, by definition of α_t, the parallel transport in N of $F_t(X)$ from $\phi_t(X)$ along the curve $t \mapsto \phi_t(X)$ to $\phi_0(X)$, i.e., for $W \in T_X M$, $(\alpha_t \cdot F_t)(X) \cdot W$ is the parallel transport in N of the vectors $F_t(X) \cdot W \in T_{\phi_t(X)}N$ along the curve $c(t) = \phi_t(X) \in N$. Now in coordinate charts we use the standard formula

$$\frac{d}{dt}(\alpha_t)^a{}_b\Big|_{t=0} = \gamma^a{}_{cb}U^c.$$

Thus

$$\frac{d}{dt}(\alpha_t \cdot F_t)^a{}_A(X)\Big|_{t=0} = \gamma^a{}_{cb}U^c(\overset{\circ}{F})^b{}_A + \frac{d}{dt}\left(\frac{\partial\phi_t{}^a}{\partial X^A}\right)\Big|_{t=0}$$

$$= \gamma^a{}_{cb}U^c(\overset{\circ}{F})^b{}_A + \frac{\partial U^a}{\partial X^A} = U^a{}_{|A}.$$

<u>Proposition</u> The linearized equations at $\overset{\circ}{\phi}$ for the map $\phi \longmapsto J$ = Jacobian of ϕ are

$$L(\overset{\circ}{\phi},V) = \overset{\circ}{J} + \overset{\circ}{J}\,\mathrm{div}\, v$$

where $\overset{\circ}{J}$ is the Jacobian of $\overset{\circ}{\phi}$ and $v = V \circ \overset{\circ}{\phi}{}^{-1}$.

<u>Proof</u>. Since the scalars over maps form a trivial bundle, i.e., a linear space, the definition gives

$$L(\overset{\circ}{\phi},V) = \overset{\circ}{J} + \frac{d}{dt}J(\phi_t)\Big|_{t=0},$$

where V is tangent to ϕ_t at $t=0$. From Section 1, this is $\overset{\circ}{J} + \overset{\circ}{J}\,\mathrm{div}\, v$ where v is the corresponding spatial velocity. □

<u>Theorem</u> Let $\mathscr{K}$ denote the space of all C^k regular configurations $\phi : M \longrightarrow N$, let $\mathscr{E}$ denote the bundle of C^{k-1} two point tensors F over $\mathscr{K}$ as above and let $\hat{T}$ map two point tensors F pointwise to two point tensors $\hat{T}(F)$ (perhaps of a different rank) and be C^r, $r \geq k$. Let $\mathscr{F}$ denote a bundle of C^{k-1} two-point tensor fields over $\mathscr{K}$ containing the range of $\hat{T}$ and let

$$f : \mathscr{K} \longrightarrow \mathscr{F};\quad \phi \longmapsto \hat{T} \circ F,\quad F = T\phi.$$

Then f is C^r and the linearized equations at $\overset{\circ}{\phi}$ are

$$L(\overset{\circ}{\phi},V) = \overset{\circ}{\hat{T}}\circ\overset{\circ}{F} + \frac{\partial\overset{\circ}{\hat{T}}}{\partial F}\cdot\nabla V$$

where $T = \hat{T}(\overset{\circ}{F})$, etc.

This is a covariant version of the composition theorem given earlier. It is proved by combining the argument given there with that for the linearization of the map $f:\phi \longmapsto T\phi = F$ given above.

In particular we can apply this result in the case when $\hat{T}$ is the first Piola-Kirchhoff stress tensor (with a given constitutive dependence on F assumed). Then we have

$$L(\overset{\circ}{\phi},V) = \overset{\circ}{\hat{T}}(\overset{\circ}{F}) + \overset{\circ}{A}\cdot\nabla V$$

where, from Section 3,

$$\overset{\circ}{A} = \frac{\partial\overset{\circ}{\hat{T}}}{\partial F}, \qquad \text{i.e.,} \quad \overset{\circ}{A}{}^{ABa}{}_{b} = \frac{\partial\hat{T}^{Aa}}{\partial F^{b}{}_{B}}(\overset{\circ}{F}),$$

is the elasticity tensor at $\overset{\circ}{\phi}$. Also,

$$(A\cdot\nabla V)^{Aa} = A^{ABa}{}_{b}V^{b}{}_{|B}.$$

<u>Example</u> The linearized equations of elastostatics; viz.,

$$\mathrm{DIV}\,T + \rho_0 B = 0,$$

for B given† at a configuration $\overset{\circ}{\phi}$ are

$$\mathrm{DIV}\,\overset{\circ}{T} + \rho_0\overset{\circ}{B} + \mathrm{DIV}(\overset{\circ}{A}\cdot\nabla V) = 0,$$

---

†Strictly speaking, B "given" requires N to be a linear space to make sense. Similarly for "prescribed" tractions τ on ∂M. Given B or τ as functions of X is usually called "dead loading". If $b = B\circ\phi$ is given then extra terms $\nabla b\cdot V$ must be added to the linearized equations.

i.e., $\mathrm{DIV}[\overset{\circ}{T} + (P \otimes G^{\sharp} + 2\mathbb{C})\cdot F\cdot\nabla V] + \rho_0 B = 0,$

where $\overset{\circ}{T} = \hat{T}(\overset{\circ}{F})$ and $\overset{\circ}{A}$ is the elasticity tensor evaluated at the configuration $\overset{\circ}{\phi}$. Recalling that, from the Piola identity

$$\mathrm{DIV}(\overset{\circ}{A}\cdot\nabla V) = J\,\mathrm{div}\,(\overset{\circ}{a}\cdot\nabla v)\circ\overset{\circ}{\phi},$$

the linearized equations may be written in spatial coordinates

$$\overset{\circ}{\rho}b + \mathrm{div}(\overset{\circ}{t} + \overset{\circ}{a}\cdot\nabla v) = 0,$$

or, equivalently,

$$\overset{\circ}{\rho}b + \mathrm{div}(\overset{\circ}{t} + [\overset{\circ}{t}\otimes\delta + \overset{\circ}{\mathbb{c}}]\cdot\nabla v) = 0.$$

Notice that if $\overset{\circ}{\phi}$ is a <u>stress free state</u> in equilibrium, i.e., $\overset{\circ}{t} = 0$, then these reduce to the equations of <u>classical linear elastostatics</u>:

$$\rho b + \mathrm{div}(\mathbb{c}\cdot\nabla v) = 0,$$

i.e., $\rho b^a + (c^{abcd} v_{c|d})_{|b} = 0,$

i.e., $\rho b^a + (c^{abcd}\varepsilon_{cd})_{|b} = 0,$

where $\varepsilon_{cd} = \frac{1}{2}(v_{c|d} + v_{d|c}) = \frac{1}{2}\mathcal{L}_v g$, in view of the symmetries of $\mathbb{c}$.

For practice the reader may wish to show that the linearization of $F^a{}_A T^{Ab} = F^b{}_A T^{Aa}$ (balance of moment of momentum) gives symmetry of

$$J[\overset{\circ}{t}\cdot\nabla v + \overset{\circ}{a}\cdot\nabla v].$$

If we consider the space of motions in place of the space of configurations we can derive in the same way the linearization of the equations of motion.

Theorem   The linearization of the equations of motion

$$\rho_0 A = \mathrm{DIV}\, T + \rho_0 B,$$

at a motion $\overset{\circ}{\phi}_t$ are

$$\rho_0(\overset{\circ}{A} + \ddot{U} - B) = \mathrm{DIV}(\overset{\circ}{T} + \overset{\circ}{A}\ \nabla U),$$

where U is a vector field over the motion $\overset{\circ}{\phi}_t$.

In spatial coordinates, these read

$$\rho(\overset{\circ}{a} + \ddot{u} - b) = \mathrm{div}(\overset{\circ}{t}^1 + \overset{\circ}{a}\cdot\nabla u).$$

The terms $-\rho_0\overset{\circ}{A} + \rho_0 B + \mathrm{DIV}\,\overset{\circ}{T}$ which measure how close $\overset{\circ}{\phi}_t$ is to an actual solution are called the out-of-balance forces. (They arise naturally when a nonlinear problem is solved by iteration on a linearized problem.)

The boundary conditions can be linearized by following the same procedures. We can state the results as follows:

Proposition   The linearization of the boundary condition of place, viz.,

$$\phi = g \text{ given on } \partial M$$

at a configuration ϕ^1 is

$$L(\phi^1, U) = \phi^1 - g + U = 0 \text{ on } \partial M.$$

(For this to make sense, we assume $N = \mathbb{R}^n$.)

The linearization of the boundary condition of traction, viz.,

$$\langle N, T\rangle = \tau \text{ given on } \partial M$$

is

$$N^1\cdot A^1\cdot\nabla U = \tau - \langle N^1, T^1\rangle \text{ on } \partial M.$$

Here, as usual, U is a vector field over ϕ^1.

Finally, the linearization of the initial conditions ($\phi_0 = d$ given) and ($V_0 = V$ given) yield

$$\phi_0^1 - d + U_0 = 0 \quad \text{and} \quad V_0^1 - V + U_0 = 0.$$

Now that we have derived the linear theory, we make a few comments to enable the reader to connect this with the classical linear theory presented by Gurtin [106].

We can summarize what we have done by saying that if we linearize about a stress free stationary undeformed state, the equations of motion are

$$\rho\ddot{u} = \rho b + \operatorname{div}(\mathbb{c}\cdot\nabla u).$$

The state ϕ^1 being undeformed means that $\phi^1 = id$ and hence there is no difference between spatial and material coordinates.

The linearization of the Lagrangian strain tensor, i.e., $E = \frac{1}{2}(C - G)$ yields the infinitesimal strain tensor

$$e = \tfrac{1}{2}\mathcal{L}_u g,$$

i.e.,

$$e_{ab} = \tfrac{1}{2}(u_{a|b} + u_{b|a}).$$

We also set

$$s = \mathbb{c}\cdot\nabla u \quad \text{or} \quad s^{ab} = c^{abcd}u_{c|d},$$

i.e.,

$$s = \mathbb{c}\cdot e \quad \text{or} \quad s^{ab} = c^{abcd}e_{cd},$$

and call s the <u>linearized stress tensor</u>. The fact that s depends only on e is the infinitesimal analogue of objectivity or material frame indifference in the finite theory. As in the finite theory, symmetry of $\mathbb{c}$

is equivalent to the existence of a stored energy function. Here it is given by

$$\varepsilon = \tfrac{1}{2} e \cdot \mathbb{C} \cdot e$$

$$= \tfrac{1}{2} e_{ab} C^{abcd} e_{cd}$$

and

$$s = \frac{\partial \varepsilon}{\partial e}, \qquad \mathbb{C} = \frac{\partial^2 \varepsilon}{\partial e \partial e}$$

as in the finite theory.

In the isotropic case (for a detailed study of more general symmetries, see Gurtin's article), we saw in Section 3 that the internal energy function was a function only of the invariants of C. We worked out P and $\mathbb{C}$ in that case. For isotropic linear elasticity we similarly conclude that ε depends only on the invariants of e. Since s is linear in e, the stress strain relationship must have the form

$$s = \lambda(\mathrm{tr}\, e) g^{\sharp} + 2\mu e^{\sharp},$$

for constants λ and μ. (This follows from our earlier formula relating P and Ψ in the isotropic case, after linearization.) The corresponding ε may therefore be taken to be

$$\varepsilon = \mu e \cdot e + \frac{\lambda}{2}(\mathrm{tr}\, e)^2.$$

Thus, in this case, the elasticity tensor is determined by the two constants λ and μ, called the <u>Lamé moduli</u>. Explicitly,

$$C^{abcd} = \lambda g^{ab} g^{cd} + 2\mu g^{ac} g^{bd}.$$

Note that the mean stress is

$(1/3)\mathrm{tr}\, s = (\lambda + (2/3)\mu)\cdot \mathrm{div}\, u.$

Thus, one calls $k = (3\lambda+2\mu)/3$ the modulus of compression and $1/k$ the compressibility. Other constants of note are

$$\beta = \mu(3\lambda+2\mu)/(\lambda+\mu) \quad \text{(Young's modulus)},$$

and

$$\nu = \lambda/2(\mu+\lambda) \quad \text{(Poisson's ratio)}.$$

These constants arise by considering the e corresponding to various simple types of s. For details of these and other constants we refer to Gurtin's article and the classical texts.

We shall, for the remainder of this section, discuss in general terms the relationship between the linearized and nonlinear theories. After we have developed the necessary tools, we shall apply these ideas to elasticity in Sections 8 and 11.

First of all, if a linearized theory has an existence and uniqueness theorem associated with it, we can conclude that the nonlinear problem has unique solutions for nearby data. That this is the case is the content of the inverse function theorem, whose statement we now recall (see, e.g., Dieudonné [68] for the proof).

Theorem (Inverse Function Theorem) Let X, Y be Banach spaces, $U \subset X$ open and $f : U \subset X \to Y$ of class C^r, $r \geq 1$. Let $x_0 \in U$ and assume $Df(x_0)$ is a linear isomorphism of X onto Y.

Then there exists (open) neighborhoods U_0 of x_0 and V_0 of $f(x_0)$ such that f maps U_0 bijectively (1-1 and onto) to V_0 and has a C^r inverse $f^{-1} : V_0 \to U_0$.

As in advanced calculus, we can derive from this, the implicit function theorem. The idea of using this result (or something equivalent to it) in nonlinear partial differential equations is an old one, going back at least to Banach himself. (See, e.g., Morrey [191], §6.3, and Fischer and Marsden [87] for some applications.)

In elasticity these results may be applied as follows. Consider the displacement problem† for elastostatics and suppose $\mathring{\phi}$ is a solution which is, say, stress free. Thus, $\mathring{\phi}$ satisfies an equation of the form $f(\mathring{\phi}) = 0$ where f is a map including the equations for elastostatics and the boundary conditions. If the loads b and boundary conditions d are small and if the linearized problem has unique solutions, the inverse function theorem guarantees that $f(\phi) = (b,d)$ has unique solutions. This is, basically, the method used by Stoppelli [235] in his analysis of the displacement and traction problems. We return to these in Sections 8 and 11.

Globally, we do not expect solutions to the displacement problem to be unique. That is, for large displacements d on the boundary and/or large forces b we expect non-uniqueness due to buckling, i.e., bifurcations. For situations or regions in which one may have uniqueness, the <u>continuity method</u>, used to prove the following result may be useful.*

<u>Proposition</u> Suppose that $f : X \longrightarrow Y$ is C^1 and that at each point $x \in X$, $Df(x)$ is an isomorphism. Suppose there is a constant $m > 0$ such that $\|Df(x)\| \geqslant m$ for all $x \in X$. Then f is a C^1 diffeomorphism of X <u>onto</u> Y.

<u>Proof</u>. First we prove f is onto. Let $x_0 \in X$ and $y_0 = f(x_0)$. Let

---

†In buckling problems, $Df(x)$ will become noninvertible at the point of buckling; i.e., the bifurcation point.

*This comment is based on some remarks of M. Gurtin and the (1978) thesis of S. Spector at Carnegie-Mellon University.

$y_1 \in Y$. We join y_0 to y_1 with a straight line by means of the map $\sigma(t) = ty_0 + (1-t)y_1$. For small t, $\sigma(t)$ lies in a neighbourhood of y_0 so there is a unique $\rho(t)$ in a neighbourhood of x_0 such that $f(\rho(t)) = \sigma(t)$. Consider continuous curves emanating from x_0 which map to $\sigma(t)$. They are locally unique and locally exist, so just as in ordinary differential equations they have a maximum domain of extendability, say $0 \leqslant t < t_0$ and are globally unique on this domain. We want to prove that $t_0 = 1$. Indeed, it is enough to show that $\rho(t_n)$ converges as $t_n \to t_0$, for then by local extendability, $\rho(t)$ could be extended beyond t_0, so t_0 would have to be 1. However we have, by the chain rule,

$$Df(\rho(t))\rho'(t) = \sigma'(t) = y_0 - y_1,$$

and so

$$\rho'(t) = Df(\rho(t))^{-1}(y_0 - y_1),$$

and hence

$$\|\rho'(t)\| \leqslant m\|y_0 - y_1\|.$$

Thus,

$$\|\rho(t_n) - \rho(t_m)\| \leqslant m\|y_0 - y_1\| \, |t_n - t_m|,$$

so $\rho(t_n)$ is a Cauchy sequence and hence converges. (Here the uniformity of $\|Df^{-1}\|$ is used in a crucial way.)

Next we prove that f is one-to-one. This requires a topological argument (indeed the result is not true for maps between riemannian manifolds, $f : \mathbb{R} \to S^1$, $x \mapsto e^{ix}$ being a case in point). The topological property used is simple connectivity of $\mathbb{F}$, defined below. Suppose that $f(x_1) = f(x_2)$. Without loss of generality, we can suppose $x_1 = 0$ and $f(0) = f(x_2) = 0$, after a translation on range and target is performed. Let $\sigma : [0,1] \times [0,1] \to \mathbb{F}$ be defined by

$$\sigma(\lambda,t) = \lambda f(tx_2).$$

For each fixed λ, $\sigma_\lambda(t) = \lambda f(tx_2)$ is a closed curve in $\mathbb{F}$ from 0 at $t=0$ to 0 at $t=1$. (Fig. 7)

Fig. 7.

As λ decreases to 0 we get a shrinking family of closed orbits. (σ is called a homotopy.)

The argument given above in the onto part of the proof actually proves the following <u>path lifting property</u>: if $\sigma(t) \in Y$ is a C^1 path in Y, $\sigma(0) = y_0$ and $f(x_0) = y_0$, there is a unique C^1 path $\rho(t)$ in x with $\rho(0) = x_0$ and $f(\rho(t)) = \sigma(t)$.

Now lift, uniquely, each of the paths $\sigma_\lambda(t)$ to obtain paths $\rho_\lambda(t)$ with $\rho_\lambda(0) = 0$ and $f(\rho_\lambda(t)) = \sigma_\lambda(t)$. Observe that $\rho_1(1) = x_2$ since $\rho_1(t) = tx_2$ is the unique path mapped to $\sigma_1(t) = f(tx_2)$ by f.

However, there are neighborhoods U of x_2 and V of 0 such that $x \in U$, $f(x) = 0$ implies $x = x_2$, again by local invertibility of f, i.e., the inverse function theorem. On the other hand, $\rho_\lambda(1)$ is a continuous function of λ, so $\{\lambda \mid \rho_\lambda(1) = x_2\}$ is a nonempty open set. Being closed as well, this set must equal $[0,1]$ by connectedness. Hence $\rho_0(1) = x_2$. But $\rho_0(t) = 0$ since $\sigma_0(t) = 0$. Thus $x_2 = 0$, so f is one to one. □

As we said in the introduction, in many situations the inverse function theorem cannot be applied and then it becomes more difficult to determine the structure of the space of solutions of the non-linear problem†. For this, methods of bifurcation theory (or singularities of mappings) are useful. Towards this end we shall now describe a few concepts from the related theory of linearization stability (see Fischer and Marsden [87]).

Let X, Y be Banach spaces and $f : U \subset X \longrightarrow Y$ be a C^1 map. We may be interested in solving $f(x) = 0$ for $x \in X$. Suppose $x_0 \in X$ is a given solution: $f(x_0) = 0$. Then, the linearized equations are simply

$$Df(x_0)\cdot h = 0.$$

Definition We say f is linearization stable at x_0 if for every solution h of the linearized equations there exists a C^1 curve $x(\varepsilon) \in X$ defined for ε in some half open interval $0 \leqslant \varepsilon < \varepsilon_0$ with $x(0) = x_0$, $f(x(\varepsilon)) = 0$ and $x'(0) = h$.

Another way of putting this is: we desire to complete h to a full solution in a perturbation expansion

$$x(\varepsilon) = x_0 + \varepsilon h + \varepsilon^2 h^2 \quad + \cdots , \qquad \varepsilon \geqslant 0$$

Of course, this is not exactly the definition since we did not require $x(\varepsilon)$ to be analytic in ε but only C^1. However this is a technical point which can be adjusted to suit the situation.

We shall also speak of h as an infinitesimal deformation of the equations $f(x) = 0$ and of a curve $x(\varepsilon)$ of exact solutions through x_0 as

---

†For elastodynamics (without dissipative mechanisms) the inverse function theorem also fails to yield local solutions to the initial value problem. See Section 11.

an actual, or finite deformation. Thus linearization stability can be phrased this way: every infinitesimal deformation is tangent to a finite deformation. We also say that infinitesimal deformations which are tangent to finite deformations are integrable. Of course, if the above definition fails, we say f is linearization unstable at x_0.

Theorem Let $f : U \subset X \longrightarrow Y$ be C^1 and $f(x_0) = 0$. Assume $Df(x_0)$ is surjective and its kernel splits. Then f is linearization stable at x_0.

Proof. Write $X_1 = \ker Df(x_0)$ and $X = X_1 \times X_2$. Then $D_2f(x_0) : X_2 \longrightarrow Y$ is an isomorphism. Thus, by the implicit function theorem, near x_0, the equation

$$f(x_1,x_2) = 0$$

can be solved for a C^1 function $x_2 = g(x_1)$. Let $x_0 = (x_{01},x_{02})$. Thus $g(x_{01}) = x_{02}$ and by implicit differentiation,

$$D_1f(x_{01},x_{02}) + D_2f(x_{01},x_{02}) \circ Dg(x_{01}) = 0.$$

Thus, $Dg(x_{01})$ is zero on X_1.

Let

$$x(\varepsilon) = (x_{01} + \varepsilon h, \ g(x_{01} + \varepsilon h)),$$

which makes sense since $h \in X$. Clearly $x(\varepsilon)$ is a C^1 curve and $x(0) = x_0$. Also

$$x'(0) = (h, Dg(x_{01}) \cdot h) = (h,0) = h$$

since $Dg(x_{01}) \cdot h = 0$. □

If f is C^r or analytic, so is $x(\varepsilon)$. However, see example (c) below.

While the conditions of this Theorem are sufficient they are not always necessary. However, in some important cases they can be shown also to be necessary†.

Let us now consider three simple examples to clarify the sort of things that can happen (see Fig. 8).

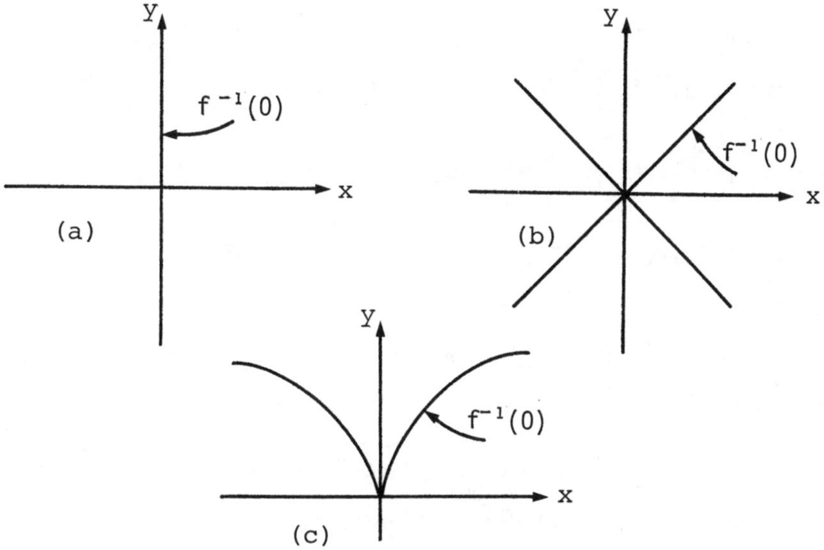

Fig. 8.

Examples (a) Let $f : \mathbb{R}^2 \longrightarrow \mathbb{R}$, $f(x,y) = x(x^2 + y^2)$. Then $f^{-1}(0) = \{(x,y) \mid f(x,y) = 0\}$ is the y-axis and $Df(0,0) = 0$. f is linearization unstable at $(0,0)$. Indeed, a vector (h_1,h_2) is integrable if and only if $h_1 = 0$, i.e., is tangent to the y-axis, although any (h_1,h_2) is an infinitesimal deformation of $f(x,y) = 0$.

(b) Let $f : \mathbb{R}^2 \longrightarrow \mathbb{R}$, $f(x,y) = x^2 - y^2$. Again linearization stability at $(0,0)$ fails. $f^{-1}(0) = \{(x,y) \mid x = \pm y\}$,

---

†For some conditions under which necessity holds, see Bourguignon, Ebin and Marsden [32]. A trivial example showing that the conditions are not always necessary is given by $f(x,y) = (x,0)$, $f : \mathbb{R}^2 \longrightarrow \mathbb{R}^2$. A more sophisticated example is the scalar curvature equation on two manifolds. See Fischer and Marsden [86]. Other conditions under which necessity holds can be given in terms of bifurcation theory.

which is not a manifold. An infinitesimal deformation (h_1, h_2) is integrable if and only if $h_1 = \pm h_2$.

(c) Let $f : \mathbb{R}^2 \longrightarrow \mathbb{R}$, $f(x,y) = x^2 - y^3$. Here the origin is again linearization unstable. The integrable directions are along the positive y-axis. The curves of exact solutions in the direction of $(0,h_2)$, $h_2 > 0$ are given by

$$(x(\varepsilon), y(\varepsilon)) = (\pm\varepsilon^{3/2}, \varepsilon h), \quad \varepsilon > 0.$$

Notice that $(x(\varepsilon), y(\varepsilon))$ is C^1 and although not analytic in ε, is, however, analytic in $\sqrt{\varepsilon}$. This sort of phenomena is an important feature of linearization instability (and occurs, for instance, in the traction problem).

Next we shall derive some necessary second order conditions which must be satisfied if an infinitesimal deformation is integrable. (For the traction problem these become the Signorini compatibility conditions ... see Section 11.)

Theorem (Compatability Conditions) Assume $f : U \subset X \longrightarrow Y$ is C^2, $h \in \ker Df(x_0)$ and that h is integrable to a C^2 finite deformation, $x(\varepsilon)$. If $\ell \in Y^*$ is orthogonal to the range of $Df(x_0)$, i.e., $\ell(Df(x_0) \cdot u) = 0$ for all $u \in X$,† then h must satisfy

$$\ell(D^2 f(x_0) \cdot (h,h)) = 0.$$

Proof. Differentiation of $f(x(\varepsilon)) = 0$ gives

$$Df(x(\varepsilon)) \cdot x'(\varepsilon) = 0.$$

---

†Another way of saying this is: $\ell \subset \ker(Df(x_0)^*)$. In Section 6 we shall see how to compute this condition for concrete elliptic operators.

On differentiating again, we have

$$D^2f(x(\varepsilon))\cdot(x'(\varepsilon),x'(\varepsilon)) + Df(x(\varepsilon))\cdot x''(\varepsilon) = 0.$$

Setting $\varepsilon = 0$:

$$D^2f(x_0)\cdot(h,h) + Df(x_0)\cdot x''(0) = 0.$$

Applying ℓ then gives the result since ℓ applied to the second term is zero. □

Likewise we can develop conditions of the third and higher order. For instance, if ℓ is orthogonal to the ranges of $Df(x_0)$ and $D^2f(x_0)$ then we must have

$$\ell(D^3f(x_0)\cdot(h,h,h)) = 0,$$

if h is to be integrable to a C^3 finite deformation.

In examples (a) and (b) above, the third and second order conditions respectively pick out the directions of linearization stability. In example (c), if we rescale to make f homogeneous by considering instead

$$F(x,y,\lambda) = f(\sqrt{\lambda}x,y) = \lambda x^2 - y^3, \quad \lambda > 0,$$

then the third order condition on F yields the directions of linearization stability. This method of rescaling will recur in our description of the traction problem in Section 11.

Conditions which are sufficient to guarantee linearization stability involve rescaling and bifurcation theory.

5 SEMIGROUP THEORY

This section gives an account of those parts of semigroup theory relevant for hyperbolic problems and in particular for elasticity. We shall also give some discussion of classical parabolic and hyperbolic second order equations as well as symmetric hyperbolic systems, but the main results on elasticity are given in Section 7. The theory of analytic semigroups has been omitted, since it is not required for most of our applications.

For further details on semigroup theory, see Hille-Phillips [125], Yosida [262], Kato [136] and Pazy [211].

The basic definition of a semigroup expresses, under the mildest possible assumptions, the idea that a linear evolution equation

$$\frac{du}{dt} = Au,$$

where A is an operator in a Banach space X, posesses, for initial data in the domain of A, unique solutions and that the solutions vary continuously in X as the initial data varies in the X topology, i.e., that the equations are <u>well-posed</u>.

Of course if A is a bounded operator in X, solutions are given by

$$u(t) = e^{tA}u = \sum_{k=0}^{\infty} \frac{(tA)^k}{k!} u.$$

For partial differential equations, A will in general be unbounded, so the problem is to make sense out of e^{tA}. Instead of power series, the operator analogue of the calculus formula

$$e^x = \lim_{n \to \infty} \left(1 - \frac{x}{n}\right)^{-n}$$

will turn out to be appropriate.

<u>Definitions</u> A (C_0) <u>semi-group</u> on a Banach space X is a family $U(t)$, $t \geq 0$, of bounded linear operators of X to X such that

(i) $U(t+s) = U(t) \circ U(s)$, $t,s \geq 0$, (semi-group property)

(ii) $U(0) = \mathrm{Id}$,

(iii) $U(t)x$ is t-continuous at $t = 0$ for each $x \in X$; i.e.,

$$\lim_{t \searrow 0} U(t)x = x.$$

The <u>infinitesimal generator</u> A of $U(t)$ is the (in general unbounded) linear operator given by

$$Ax = \operatorname*{limit}_{t \searrow 0} \frac{U(t)x - x}{t}$$

on the domain $D(A)$ where the limit in X exists.

We now derive a number of properties of semi-groups. (Eventually we will prove an existence and uniqueness theorem for semigroups given a generator A.) For all these properties we assume $U(t)$ is a given C^0 semigroup with infinitesimal generator A.

1. <u>There are constants</u> M, β <u>such that</u> $\|U(t)\| \leqslant Me^{t\beta}$. <u>In this case we write</u>

$$A \in G(X, M, \beta)$$

<u>and say</u> A <u>is the generator of a semigroup of type</u> (M, β).

<u>Proof</u>. We first show that $\|U(t)\|$ is bounded on some neighborhood of zero. If not, there would be a sequence $t_n \downarrow 0$ such that $\|U(t_n)\| \geqslant n$. But $U(t_n)x \longrightarrow x$ as $n \longrightarrow \infty$, so $U(t_n)$ is pointwise bounded as $n \longrightarrow \infty$, and therefore by the Banach-Steinhaus theorem (Uniform Boundedness Theorem), $\|U(t_n)\|$ is bounded, which is a contradiction.

Thus for some $\delta > 0$ there is a constant M such that $\|U(t)\| \leqslant M$ for $0 \leqslant t \leqslant \delta$. For $t \geqslant 0$ arbitrary, let n be the largest integer in t/δ so $t = n\delta + \tau$, $0 \leqslant \tau < \delta$. Then by the semigroup property,

$$\|U(t)\| = \|U(n\delta)U(\tau)\|$$

$$\leq \|U(\tau)\| \, \|U(\delta)\|^n$$

$$\leq M \cdot M^n$$

$$\leq M \cdot M^{t/\delta}$$

$$\leq Me^{t\beta},$$

where $\beta = \frac{1}{\delta}\log M$. □

2. $U(t)$ is strongly continuous[†] in t, i.e., for each $x \in X$, $U(t)x$ is continuous in X as a function of $t \in [0,\infty)$.

Proof. Let $s > 0$. Then since $U(\tau + s)x = U(s)U(\tau)x$ we obtain from (iii) in the definition,

$$\lim_{t \downarrow s} U(t)x = \lim_{\tau \downarrow 0} U(\tau + s)x$$

$$= U(s) \lim_{\tau \downarrow 0} U(\tau)x$$

$$= U(s)x,$$

so we have right continuity in t at $t = s$. For left continuity, write, for $0 \leq \tau \leq s$,

$$\|U(s - \tau)x - U(s)x\| = \|U(s - \tau)(x - U(\tau)x)\| \leq Me^{\beta(s-\tau)} \|x - U(\tau)x\|,$$

which tends to zero as $\tau \downarrow 0$. □

[†]One can show that strong continuity at $t = 0$ can be replaced by weak continuity at $t = 0$ and strong continuity in $t \in [0,\infty)$ can be replaced by strong measurability in t. See Hille-Phillips [125].

3. (i) $U(t)D(A) \subset D(A)$,

(ii) $U(t)Ax = AU(t)x$

(iii) for $x \in D(A)$, $\frac{d}{dt}U(t)\cdot x = A(U(t)x)$, $t \geqslant 0$; i.e., $x(t) = U(t)\cdot x$ satisfies

$$\frac{dx}{dt} = Ax.$$

Proof. From $\frac{1}{h}[U(h)U(t)x - U(t)x] = U(t)\left[\frac{U(h)x - x}{h}\right]$ we get (i) and (ii). We get (iii) by using the fact† that if $u(t) \in X$ has a continuous right derivative, then $u(t)$ is differentiable - from the right at $t = 0$ and two sided if $t > 0$. □

From (i) and (ii) we see that if $x \in D(A^n)$ then $U(t)x \in D(A^n)$. This is often used to derive regularity results.

4. $D(A)$ is dense in X.

Proof. Let $\phi(t)$ be a C^∞ function with compact support in $[0,\infty)$, let $x \in X$ and set

$$x_\phi = \int_0^\infty \phi(t)U(t)\cdot x\,dt.$$

Noting that

$$U(s)x_\phi = \int_0^\infty \phi(t)U(t+s)\cdot x\,dt$$

$$= \int_0^\infty \phi(\tau - s)U(\tau)x\,d\tau$$

---

†This follows from the corresponding real variables fact by considering $\ell(u(t))$ for $\ell \in X^*$. See Yosida [262], p.235.

is differentiable in s, we find that $x_\phi \in D(A)$. On the other hand, given any $\varepsilon > 0$ there is a ϕ (close to the "δ function") such that $\|x_\phi - x\| < \varepsilon$. Indeed, choose $\delta > 0$ such that $\|U(t)x - x\| < \varepsilon$ if $0 \leqslant t \leqslant \delta$. Let ϕ be C^∞ with compact support in $(0,\delta)$, $\phi \geqslant 0$ and $\int_0^\infty \phi(t)dt = 1$. Then

$$\begin{aligned} \|x_\phi - x\| &= \left\| \int_0^\delta \phi(t)(U(t)x - x)\, dt \right\| \\ &\leqslant \int_0^\delta \phi(t)\, \|U(t)x - x\|\, dt \\ &< \varepsilon \int_0^\delta \phi(t)dt = \varepsilon. \qquad \square \end{aligned}$$

This same argument in fact shows that $\bigcap_{n=1}^\infty D(A^n)$ <u>is dense in</u> X.

5. A <u>is a closed operator; i.e., its graph in</u> $X \times X$ <u>is closed</u>.†

<u>Proof</u>. Let $x_n \in D(A)$ and assume $x_n \to x_0$ $Ax_n \to y$. We must show that $x \in D(A)$ and $y = Ax$. By property 3,

$$U(t)x_n = x_n + \int_0^t U(s)Ax_n ds.$$

Since $U(s)Ax_n \to U(s)y$ uniformly for $s \in [0,t]$,

$$U(t)x = x + \int_0^t U(s)y ds.$$

It follows that $\frac{d}{dt+}U(t)x$ exists and equals y. $\square$

---

†We shall prove more than this in the proposition below, but the techniques in 4, 5, 6 apply to certain nonlinear semigroups as well. See Chernoff and Marsden [39].

6. (Existence Implies Uniqueness). Suppose $c(t)$ is a differentiable curve in X such that $c(t) \in D(A)$ and $c'(t) = A(c(t))$, $t \geq 0$. Then

$$c(t) = U(t)\cdot c(0).$$

Proof. Fix $t_0 > 0$ and define $h(t) = U(t_0 - t)c(t)$ for $0 \leq t < t_0$. Then for τ small,

$$\begin{aligned}\|h(t+\tau) - h(t)\| &= \|U(t_0 - t - \tau)c(t+\tau) - U(t_0 - t - \tau)U(\tau)c(t)\| \\ &\leq Me^{\beta(t_0 - t - \tau)}\|c(t+\tau) - U(\tau)c(\tau)\|.\end{aligned}$$

However,

$$\frac{1}{\tau}[c(t+\tau) - U(\tau)c(t)] = \frac{1}{\tau}[c(t+\tau) - c(t)] - \frac{1}{\tau}[U(\tau)c(t) - c(t)],$$

which converges, as $\tau \to 0$, to $Ac(t) - Ac(t) = 0$. Thus, $h(t)$ is differentiable for $0 < t < t_0$ with derivative zero. By continuity, $h(t_0) = \lim_{t\uparrow t_0} h(t) = c(t_0) = \lim_{t\downarrow 0} h(t) = U(t_0)c(0)$. (The last limit is justified by the fact that $\|U(t)\| \leq Me^{t\beta}$). This is the result with t replaced by t_0. □

One also has uniqueness within the class of weak solutions (see Ball [13] and Balakrishnan [10]). If $c(t)$ is a continuous curve in X and if, for every $v \in D(A^*)$, • (i.e., $\langle A^*v,w\rangle = \langle v,Aw\rangle$ for all $w \in D(A)$, where $\langle\ ,\ \rangle$ is the pairing between X and X^*), $\langle c(t),v\rangle$ is absolutely continuous and

$$\frac{d}{dt}\langle c(t),v\rangle = \langle c(t),A^*v\rangle \qquad \text{a.e.},$$

i.e., $\langle c(t),v\rangle = \langle c(0),v\rangle + \int_0^t \langle c(s),A^*v\rangle\, ds$

then $c(t)$ is called a weak solution of $dx/dt = Ax$.

7. If $c(t)$ is a weak solution, then $c(t) = U(t)c(0)$. Conversely, for $x_0 \in X$ (not necessarily in the domain of A) then $c(t) = U(t)x_0$ is a weak solution.

Proof. If $x_0 \in D(A)$ then $U(t)x_0$ is a solution in $D(A)$ and hence a weak solution. Since $U(t)$ is continuous and $D(A)$ is dense, the same is true for $x_0 \in X$, i.e., we can pass to the limit in

$$\langle U(t)x_n, v\rangle = x_n + \int_0^t \langle U(\tau)x_n, v\rangle\, d\tau$$

for $x_n \in D(A)$, $x_n \longrightarrow x_0 \in X$.

Now suppose $c(t)$ is a weak solution. Let $w(t) = c(t) - U(t)c(0)$. Then $w(0) = 0$ and for $v \in D(A^*)$,

$$\langle w(t), v\rangle = \int_0^t \langle w(\tau), A^*v\rangle\, d\tau$$

$$= \langle \int_0^t w(\tau)d\tau, A^*v\rangle .$$

Thus, $\int_0^t w(\tau)d\tau \in D(A)$ since A is closed (property 5).

Here we use the fact that if A is closed then $A^{**} \subset A$... identifying X with a subspace of X^{**}. (If X is reflexive, $A^{**} = A$; cf., Kato [136], p.168.) It follows that $z(t) = \int_0^t w(\tau)d\tau$ satisfies $\dot{z} = Az$ and since $z(0) = 0$, z is identically zero by property 6. □

Ball [13] also shows that if the equation $\dot{x} = Ax$ admits unique weak solutions and A is densely defined and closed, then A is a generator.

8. $T(t) = e^{-t\beta}U(t)$ is a uniformly bounded semigroup, i.e., $\|T(t)\| \leqslant M$, with generator $(A - \beta I)$.

This result, a simple verification, often enables one to reduce a proof to the case of bounded semigroups.

9. (Feller) *There is an equivalent norm on* X, *say* $|||\cdot|||$, *for which* $U(t)$ *is quasi-contractive, i.e.,*

$$|||U(t)||| \leqslant e^{t\beta}.$$

(*If* $\beta = 0$, *we say* U *is contractive*).

Proof. This is readily verified with

$$|||x||| = \sup_{t \geqslant 0} \| e^{-t\beta} U(t)x \|. \quad \square$$

For instance, let $X = L^2(\mathbb{R})$ with the norm

$$\|f\|^2 = \int_{-1}^{1} |f(x)|^2 dx + \tfrac{1}{2} \int_{\mathbb{R}-[-1,1]} |f(x)|^2 dx$$

and let $(U(t)f)(x) = f(t+x)$. Then $U(t)$ is a C^0 semigroup and $Af = df/dx$ with domain $H^1(\mathbb{R})$ (absolutely continuous functions with derivatives in L^2). Here,

$$\|U(t)\| \leqslant 2.$$

If we form $|||\ |||$ in property 9 we get the usual L_2 norm and a contraction semigroup.

10.† $U(t)$ *is norm continuous at* $t = 0$ *if and only if* $A \in B(X)$.

Indeed, choose $\varepsilon > 0$ so that $\|U(t) - I\| < \frac{1}{2}$ if $0 \leqslant t \leqslant \varepsilon$ and pick ϕ to be a C^∞ function with compact support in $[0,\varepsilon)$ and $\phi \geqslant 0$, $\int_0^\varepsilon \phi(t)dt = 1$.

Let $J_\phi(x) = \int_0^\infty \phi(\tau)U(\tau)x\,d\tau$ and note that

---

†See Chernoff-Marsden [39], p.62 for a nonlinear generalization.

$$J_\phi(U(t)x) = \int_0^\infty \phi(\tau)U(\tau + t)x\,d\tau$$

$$= \int_t^\infty \phi(\tau - t)U(\tau)x\,d\tau.$$

But

$$\| (J_\phi - I)(x) \| = \| \int_0^\infty \phi(\tau)(U(\tau)x - x)d\tau \|$$

$$\leqslant \tfrac{1}{2} \int_0^\infty \phi(\tau) \|x\| d\tau$$

$$= \tfrac{1}{2} \|x\|,$$

so $\|J_\phi - I\| \leqslant \frac{1}{2}$ and hence J_ϕ is invertible. But then

$$U(t)x = J_\phi^{-1}\left(\int_t^\infty \phi(\tau - t)U(\tau)x\,d\tau\right),$$

which is differentiable in t for all x and shows $A \in B(X)$. The converse is done by noting that $e^{tA} = \sum_{n=0}^{\infty}(tA)^n/(n!)$ (by uniqueness) is norm continuous in t. □

Next we give a proposition which will turn out to be a complete characterization of generators.

Proposition Let $A \in G(X,M,\beta)$. Then

(i) $D(A)$ is dense,

(ii) $(\lambda - A)$ is one to one and onto X for $\lambda > \beta$ and $R_\lambda = (\lambda - A)^{-1} \in B(X)$, ($R_\lambda$ is called the resolvent),

(iii) $\|(\lambda - A)^{-n}\| \leqslant M/(\lambda - \beta)^n$ for $\lambda > \beta$ and $n = 1,2,\ldots$

Proof. Given $x \in X$, let

$$y = \int_0^\infty e^{-\lambda t} U(t)x \, dt, \quad \lambda > \beta.$$

Then

$$U(s)y - y = \int_0^\infty e^{-\lambda t} U(t+s)x \, dt - y$$

$$= e^{\lambda s} \int_s^\infty e^{-\lambda \tau} U(\tau)x \, d\tau - y$$

$$= (e^{\lambda s} - 1)y - e^{\lambda s} \int_0^s e^{-\lambda t} U(t)x \, dt.$$

Hence $y \in D(\dot{A})$ and $Ay = \lambda y - x$. Thus $(\lambda - A)$ is surjective. (Taking $\lambda \longrightarrow \infty$ shows $\lambda y \longrightarrow x$ which also shows $D(A)$ is dense.) The formula

$$u = \int_0^\infty e^{-\lambda t} U(t)(\lambda - A)u \, dt, \quad u \in D(A),$$

which follows from $-\frac{d}{dt} e^{-\lambda t} U(t)u = e^{-\lambda t} U(t)(\lambda - A)u$ shows that $(\lambda - A)$ is one to one.

Thus we have proved the <u>Laplace transform relation</u>

$$R_\lambda x = (\lambda - A)^{-1} x = \int_0^\infty e^{-\lambda t} U(t)x \, dt, \quad \lambda > \beta,$$

from which it follows that

$$\| (\lambda - A)^{-1} \| \leq \int_0^\infty e^{-\lambda t} M e^{\beta t} dt = M/(\lambda - \beta).$$

The estimate (iii) follows from the formulae

$$(n-1)!(\lambda - A)^{-n} x = \int_0^\infty e^{-\lambda t} t^{n-1} U(t)x \, dt,$$

$$\int_0^{\infty} e^{-\mu t} t^{n-1} dt = (n-1)!/\mu^n.$$

The second of these is proved by integration by parts and the first follows from what we have proved together with the relation

$$\left(\frac{d}{d\lambda}\right)^{n-1}(\lambda - A)^{-1} = (-1)^{n-1}(n-1)!(\lambda - A)^{-n}. \quad \square$$

The reader may wish to check that the <u>resolvent identity</u>

$$R_\lambda - R_\mu = (\mu - \lambda)R_\lambda R_\mu$$

holds and that $\lambda R_\lambda \longrightarrow$ Identity strongly as $\lambda \longrightarrow \infty$.

The Hille-Yosida theorem asserts the converse. It is, in effect, an existence and uniqueness theorem (see property 6 above for uniqueness).

<u>Theorem</u> Let A be an (unbounded) operator in X. Assume there are constants M, β such that

(i) $D(A)$ is dense,

(ii) $(\lambda - A)$ is one to one and onto X for $\lambda > \beta$ and $(\lambda - A)^{-1} \in B(X)$,

(iii) $\|(\lambda - A)^{-n}\| \leqslant M/(\lambda - \beta)^n$, $\lambda > \beta$, $n = 1,2,\ldots$

Then $A \in G(X,M,\beta)$; i.e., there exists a C^0 semigroup of type (M,β) whose generator is A.

<u>Remarks</u> 1 If (ii), (iii) hold for $|\lambda| > \beta$ then $U(t)$ is a <u>group</u>, i.e., is defined for all $t \in \mathbb{R}$, not just $t \geqslant 0$.

2 As the proof will show, it suffices to verify (ii) and (iii) for some sufficiently large λ.

<u>Proof</u>. If $(A - \beta)$ generates the semigroup U_t then A generates the

semi-group $e^{t\beta}U_t$ (property 8). Thus it suffices to prove the theorem for $\beta = 0$.

Rewrite (iii) as

$$\|(1-\alpha A)^{-n}\| \leq M, \quad \alpha > 0, \quad n = 1, \ldots,$$

by taking $\alpha = 1/\lambda$. Now if $x \in D(A)$,

$$(1-\alpha A)^{-1}x - x = \alpha(1-\alpha A)^{-1}Ax,$$

so $(1-\alpha A)^{-1} \longrightarrow 0$ strongly on $D(A)$ and, being uniformly bounded, on X as $\alpha \downarrow 0$.

Let $U_n(t) = (1-\frac{t}{n}A)^{-n}$, uniformly bounded. We show they converge on a dense set. Write

$$U_n(t)x - U_m(t)x = U_m(t-s)U_n(s)x \Big|_{s=0}^{s=t}$$

$$= \operatorname*{s-lim}_{\varepsilon\downarrow 0} \int_{\varepsilon}^{t} \frac{d}{ds} U_m(t-s)U_n(s)x\, ds$$

$$= \operatorname*{s-lim}_{\varepsilon\downarrow 0} \int_{\varepsilon}^{t} \left(\frac{s}{n} - \frac{t-s}{m}\right)A^2\left(1-\frac{t-s}{m}A\right)^{-m-1}\left(1-\frac{s}{n}A\right)^{-n-1}x\, ds.$$

Thus, if $x \in D(A^2)$ we get,

$$\|U_n(t)x - U_m(t)x\| \leq M^2\|A^2x\| \; \tfrac{1}{2}\left(\frac{1}{n} + \frac{1}{m}\right)t^2.$$

Thus $U_n(t)x$ converges for $x \in D(A^2)$. But

$$D(A^2) = D((1-A)^2)$$

$$= R((1-A)^{-2})$$

$$= (1-A)^{-1}D(A).$$

Now $(1-A)^{-1} : X \to D(A)$ is bounded and maps onto $D(A)$. Since $D(A) \subset X$ is dense, $(1-A)^{-1}(D(A)) \subset D(A)$ is dense, i.e., $D(A^2) \subset X$ is dense.

Let $U(t)x = \underset{n\to\infty}{\text{s-lim}}\, U_n(t)$. Clearly, $\|U(t)\| \leqslant M$, $U(0)x = x$ and $U(t+s) = U(t) \circ U(s)$. Since $U_n(t)x \to U(t)x$ uniformly on compact t-intervals for $x \in D(A^2)$ and this is dense, $U(t)x$ is t-continuous. So we have a C^0 semigroup.

Let A' be the generator of $U(t)$. We need to show that $A' = A$. For $x \in D(A)$,

$$\frac{d}{dt} U_n(t)x = A(1 - \tfrac{t}{n}A)^{-1} U_n(t)x.$$

Thus

$$U_n(t)x = x + \int_0^t (1 - \tfrac{s}{n}A)^{-1} U_n(s)Ax\, ds,$$

so

$$U(t)x = x + \int_0^t U(s)Ax\, ds,$$

and hence $x \in D(A')$ and $A' \supset A$.

But $(1-A')^{-1} \in B(X)$ by the previous proposition and $(1-A)^{-1} \in B(X)$, so they must agree. □

Remarks 1 In verifying the hypotheses, the possibility that $M > 1$ is a main difficulty. If $M = 1$ we have a quasi-contractive semi-group and verification of (iii) for $n = 1$ is sufficient.

2 We shall often write e^{tA} for the semigroup generated by A.

For applications, there are two special versions of the Hille-Yosida theorem which are particularly convenient.

First Corollary A linear operator A on X has a closure $\bar{A}$ which is the generator of a quasi-contractive semigroup on X if and only if

(i) $D(A)$ is dense,

(ii) for λ sufficiently large $(\lambda - A)$ has dense range and

$$\|(\lambda - A)x\| \geqslant (\lambda - \beta)\|x\|.$$

Proof. Necessity follows easily from the preceeding proposition. For sufficiency, we use the following.

Lemma (a) Let B be a closable linear operator with a densely defined bounded inverse B^{-1}. Then $\overline{(B^{-1})}$ is injective, and $(\overline{B})^{-1} = \overline{(B^{-1})}$.

(b) Suppose that A is a densely defined linear operator such that $(\lambda - A)^{-1}$ exists, is densely defined, with a bound K/λ as $\lambda \to \infty$. Then A is closable. (Hence by part (a), $(\lambda - \overline{A})$ is invertible, with $(\lambda - \overline{A})^{-1} = \overline{(\lambda - A)^{-1}}$).

Proof. (a) $\overline{B^{-1}}$ is a bounded, everywhere-defined operator. Suppose that $\overline{B^{-1}}y = 0$. We will show that $y = 0$. Let $y_n \in R(B)$ (range of B), $y_n \to y$. Then $y_n = Bx_n$, $x_n \in D(B)$, and $\|x_n\| \leqslant \|B^{-1}\|\,\|y_n\| \to 0$. Since B is closable, we must have $y = 0$. Thus $\overline{B^{-1}}$ is injective and (a) follows.

(b) We shall first show that $\lambda R_\lambda \to I$ as $\lambda \to \infty$, where $R_\lambda = \overline{(\lambda - A)^{-1}}$ by definition. By assumption, $\|R_\lambda\| \leqslant K/\lambda$. Now pick any $x \in D(A)$. Then $x = R_\lambda(\lambda - A)x$, so $x = \lambda R_\lambda x - R_\lambda Ax$, and $\|R_\lambda Ax\| \leqslant (K/\lambda)\|Ax\| \to 0$ as $\lambda \to \infty$. Thus $\lambda R_\lambda \to I$ strongly on $D(A)$. But $D(A)$ is dense and $\|\lambda R_\lambda\| \leqslant K$ for all large λ, so $\lambda R_\lambda \to I$ on the whole of X.

To prove that A is closable we suppose $x_n \in D(A)$, $x_n \to 0$, and $Ax_n \to y$. We claim that $y = 0$. Indeed, choose a sequence $\lambda_n \to \infty$ with

$\lambda_n x_n \to 0$. Then

$$(\lambda_n - A)x_n + y \to 0.$$

Since $\|\lambda_n R_{\lambda_n}\| \leq K$, we have

$$\lambda_n R_{\lambda_n}[(\lambda_n - A)x_n + y] \to 0.$$

So,

$$\lambda_n x_n + \lambda_n R_{\lambda_n} y \to 0.$$

But $\lambda_n x_n \to 0$ and $\lambda_n R_{\lambda_n} y \to y$, so $y = 0$. □

The rest of the theorem follows immediately, since A satisfies the conditions of part (b) of the lemma, and hence $\overline{A}$ satisfies the hypothesis of the Hille-Yosida Theorem (with $M = 1$). □

Now we give a result in Hilbert space (see Lumer and Phillips [176] for the Banach space case. It proceeds in exactly the same way, using a duality map in place of the inner product). The central idea is that of a dissipative operator and will be our main tool in subsequent sections. We will sometimes refer to this result as the Lumer-Phillips theorem ; for applications we shall give in later sections, it will be the most useful.

<u>Second Corollary</u> Let A be a linear operator in a Hilbert space X. Then A has a closure $\overline{A}$ which is the generator of a quasi-contractive semigroup on X, i.e., $A \in G(X,1,\beta)$ if and only if,

(i) $D(A)$ is dense in X,

(ii) there is a $\beta \in \mathbb{R}$ such that

$$\langle Ax,x\rangle \leq \beta\langle x,x\rangle \qquad \text{for all } x \in D(A),$$

(iii) $(\lambda - A)$ has dense range for sufficiently large λ.

Proof. First suppose (i), (ii) and (iii) hold. Then

$$\langle (\lambda - A)x,x \rangle \geqslant (\lambda - \beta)\|x\|^2,$$

and so by Schwarz's inequality

$$\|(\lambda - A)x\| \geqslant (\lambda - \beta)\|x\|.$$

Thus $\overline{A} \in G(X,1,\beta)$ by the preceding corollary.

Conversely, assume $\overline{A} \in G(X,1,\beta)$. We need only show that

$$\langle \overline{A}x,x \rangle \leqslant \beta\langle x,x \rangle \qquad \text{for all} \quad x \in D(\overline{A}).$$

By property 8 we can assume $\beta = 0$ and $U(t)$ is contractive. Now

$$\langle x,U(t)x \rangle \leqslant \|x\| \, \|U(t)x\| \leqslant \|x\|^2$$

and therefore

$$\langle x,U(t)x - x \rangle \leqslant 0.$$

Dividing by t and letting $t \downarrow 0$ gives $\langle x,\overline{A}x \rangle \leqslant 0$ as desired. □

If $\langle Ax,x \rangle \leqslant 0$, we say A is dissipative. One can rephrase the condition that $(\lambda - A)$ is onto by saying that A is maximal dissipative. (See, e.g., Pazy [211] or Pazy's notes in this volume.)

Some additional useful results, given without detailed proof are as follows:

1. (Bounded Perturbations) If $A \in G(X,M,\beta)$ and $B \in B(X)$, then $A + B \in G(X,M,\beta + \|B\|M)$ (Kato [136], p.495).

2. (Trotter-Kato Theorem) If $A_n \in G(X,M,\beta)$ and $A \in G(X,M,\beta)$ and for λ

sufficiently large,

$$(\lambda - A_n)^{-1} \longrightarrow (\lambda - A)$$

strongly, then

$$e^{tA_n} \longrightarrow e^{tA}$$

strongly, uniform on bounded t-intervals (Kato [136], p.502). [If $D(A_n)$ and $D(A)$ all have a common core $Y \subset X$, i.e., A_n, A are the closures of their restrictions to Y, and $A_n \longrightarrow A$ strongly on Y, then $(\lambda - A_n)^{-1} \underset{s}{\longrightarrow} (\lambda - A)^{-1}$ from the resolvent identity.]

3. (Lax Equivalence Theorem) If $A \in G(X,M,\beta)$ and $K_\varepsilon \in B(X)$, $\varepsilon \geqslant 0$, with $K_0 = \mathrm{Id}$, we say $\{K_\varepsilon\}$ is

(i) stable if $\|K_{t/n}^n\|$ is bounded on bounded t-intervals, $n = 1,2,\ldots$,

(ii) resolvent consistent if for λ sufficiently large

$$(\lambda - A)^{-1} = \underset{\varepsilon\downarrow 0}{\text{s-limit}}\ (\lambda - \tfrac{1}{\varepsilon}(K_\varepsilon - I))^{-1},$$

(iii) consistent if $\frac{d}{d\varepsilon+} K_\varepsilon(x)|_{\varepsilon=0} = Ax$, $x \in$ a core of A.

Then $e^{tA} = \underset{n\to\infty}{\text{s-limit}}\ K_{t/n}^n$ uniformly on bounded t-intervals if and only if $\{K_\varepsilon\}$ is stable and resolvent consistent (See, Chorin, Hughes, McCracken and Marsden [46] for a proof and applications). Consistency implies resolvent consistency (assuming stability).

4. (Trotter Product Formula) If A, B are generators of quasi-contractive semigroups and $C = \overline{A + B}$ is a generator, then

$$e^{tC} = \underset{n\to\infty}{\text{s-limit}}\ (e^{tA/n}e^{tB/n})^n.$$

(This is a special case of 3.)

5. (Inhomogeneous equations) Let $A \in G(X,M,\beta)$ and consider the following initial value problem: Let $f(t)$, $0 \leqslant t \leqslant T$, be a continuous X-valued function. Find $x(t)$, $0 \leqslant t \leqslant T$, with $x(0)$ a given member of $D(A)$, such that

$$x'(t) = Ax(t) + f(t). \qquad \text{(I)}$$

If we solve (I) formally, by the variation of constants formula, we get

$$x(t) = e^{tA}x(0) + \int_0^t e^{(t-\tau)A}f(\tau)d\tau, \quad 0 \leqslant t \leqslant T.$$

$x(t)$ need not lie in $D(A)$, however; but it will if f is a C^1 function from $[0,T]$ to X. Then (I) is satisfied in the classical sense (Kato [136], p.486). For uniqueness, suppose $y(t)$ is another solution of (I), with $y(0) = x(0)$. Let $z(t) = x(t) - y(t)$. Then

$$\begin{cases} z'(t) = Az(t) \\ z(0) = 0 \end{cases}$$

and so $z(t) \equiv 0$ by property 6. Thus $x(t) = y(t)$.

6. (Trend to Equilibrium) Let $A \in G(X,1,\beta)$ and suppose there is a $\delta > 0$ such that the spectrum of (e^A) lies inside the unit disc a positive distance δ from the unit circle. Then for any $x \in X$,

$$e^{tA}x \longrightarrow 0 \quad \text{as} \quad t \longrightarrow +\infty.$$

[If $0 < \delta' < \delta$, we can, via the spectral theorem, find a new norm in which

$$A \in G(X,1,-\delta'),$$

from which the result is trivial. (See Marsden and McCracken, [185], §2A, and Slemrod [230].) See also Hille-Phillips [125] for conditions under which spectrum $e^A = e^{\text{spectrum } A}$, e.g., the spectrum of A is discrete. This result complements Liapunov techniques in, for instance, Dafermos [63].]

The following is an interesting abstract interpolation inequality.

Example (Kato [141]) If $A \in G(X,M,1)$ and $u \in D(A^2)$, then

$$\|Au\|^2 \leqslant 2M(M+1)\|u\|\|A^2u\|.$$

(For contraction semi-groups, $2M(M+1) = 4$, which may be replaced by 2 in Hilbert space.)

Proof. From $(d^2/dt^2)e^{tA}u = e^{tA}A^2u$, one gets

$$e^{tA}u = u + tAu + \int_0^t (t-s)e^{sA}A^2u\,ds,$$

from which $t\|Au\| \leqslant \|u\| + M\|u\| + M\|A^2u\| \int_0^t (t-s)ds$, i.e.,

$$\|Au\| \leqslant (1+M)\|u\|/t + \tfrac{1}{2}M\|A^2u\|t.$$

The elementary inequality $at + \frac{b}{t} \geqslant 2\sqrt{ab}$ then gives the result. □

For instance, consideration of the translation semigroup on $[0,\infty)$ in L_p gives the inequality

$$\|u'\|_{L_p} \leqslant 4\|u\|_{L_p}\|u''\|_{L_p}.$$

We continue now with a few remarks on operator theory in Hilbert space and its relationship to semigroup theory. The results, due to Stone and von Neumann, are classical and may be found in any of the aformentioned references. Recall that a densly defined operator (in Hilbert space) is

symmetric if $A \subset A^*$,

self adjoint if $A = A^*$,

essentially self adjoint if $\bar{A}$ is self adjoint.

For the first two results following, X is assumed to be a complex Hilbert space.

1. Let A be closed and symmetric. Then A is self adjoint if and only if $A + \lambda I$ is surjective when $\operatorname{Im}\lambda \neq 0$.

2. (Stone's Theorem) A is self adjoint if and only if iA generates a one parameter unitary group.

[From the symmetry of A and $\|(A+\lambda)x\|^2 \geqslant 0$ we get

$$\|(A+\lambda)x\| \geqslant |\operatorname{Im}\lambda| \, \|x\|,$$

and so for λ real,

$$\|(\lambda - iA)x\| \geqslant |\lambda| \, \|x\|.$$

Thus 2 results from 1 and the Hille-Yosida Theorem.]

3. (Real Stone's Theorem) Let A be a skew adjoint operator on a real Hilbert space (i.e., $A = -A^*$). Then A generates a one parameter group of isometries and conversely.

[This follows by an argument similar to 2.]

4. Let A be closed, symmetric and $A \leqslant 0$, i.e., $\langle Ax, x\rangle \leqslant 0$ for all x. Then A is self adjoint if and only if $(\lambda - A)$ is onto, $\lambda > 0$.

5. If A is dissipative $(\langle Ax, x\rangle \leqslant 0)$ and self adjoint then A generates a contraction semigroup.

[This follows from 4 and the Lumer-Phillips theorem.]

Example (Heat Equation) Let $\Omega \subset \mathbb{R}^n$ be an open region with smooth boundary,

$$X = L_2(\Omega),$$

$$Au = \Delta u, \qquad D(A) = C_0^\infty(\Omega),$$

where $C_0^\infty(\Omega)$ are the C^∞ functions with compact support in Ω. Then $\bar{A}$ generates a contraction semigroup in X.

Proof. Obviously A is symmetric and hence closable. Moreover, for $u \in D(A)$,

$$\langle Au, u\rangle = \int_\Omega \Delta u \cdot u \, dx$$

$$= -\int_\Omega \nabla u \cdot \nabla u \, dx \leqslant 0,$$

so A is dissipative. By the second corollary of the Hille-Yosida theorem, we must show that for $\lambda > 0$, $(\lambda - A)$ has dense range, i.e., $\bar{A}$ is self adjoint. Suppose $v \in L_2(\Omega)$ is such that

$$\langle (\lambda - A)u, v\rangle = 0 \qquad \text{for all } u \in D(A).$$

Then

$$\int_\Omega \langle (\lambda + \Delta)u, v\rangle \, dx = 0 \qquad \text{for all } u \in C_0^\infty(\Omega).$$

At this point we must use a fact about regularity of solutions of elliptic equations. The above states that $(\lambda + \Delta)v = 0$ in the sense of distributions, i.e., v is a weak solution. Then v is in fact C^∞ and

$v = 0$ on $\partial\Omega$. (Proofs of these results are found in, for example, Agmon [2], Mizohata [190] or Morrey [191] and are stated for more general operators below.) Thus, setting $u = v$ and integrating by parts, gives

$$\lambda\int_\Omega |v|^2 dx + \int_\Omega |\nabla v|^2 dx = 0,$$

and so $v = 0$. □

More careful considerations show that $\overline{A} = \Delta$ on $D(\Delta) = \{u \in H^2(\Omega) \mid u = 0 \text{ on } \partial\Omega\}$.

We can generalize this example somewhat. It is useful to do so since it involves concepts which we will need later. We continue to work with scalar equations, although we will need to eventually work with <u>systems</u>.

Again let $\Omega \subset \mathbb{R}^n$ have a smooth boundary and let $X = L_2(\Omega)$. Consider a differential operator of order 2:

$$Au = \sum_{i,j} a_{ij} \frac{\partial^2 u}{\partial x^i \partial x^j} + \sum_i b_i \frac{\partial u}{\partial x^i} + cu,$$

where a_{ij}, b_i and c are smooth functions. We can assume $a_{ij} = a_{ji}$.

Let $D(A) = \{u \in H^2(\Omega) \mid u = 0 \text{ on } \partial\Omega\}$.

<u>Definition</u> The <u>principal symbol</u> of A is

$$\sigma(x,\xi) = \sum_{i,j} a_{ij}(x)\xi^i\xi^j,$$

where $\xi \in \mathbb{R}^n$. We say A is <u>strongly elliptic</u> if there is an $\varepsilon > 0$ such that

$$\sigma(x,\xi) \geqslant \varepsilon |\xi|^2$$

for all $x \in \Omega$, $\xi \in \mathbb{R}^n$.

The Dirichlet form is defined by

$$B(u_1,u_2) = \sum_{i,j=1}^{n} \int_\Omega a_{ij}(x)\frac{\partial u_1}{\partial x^i}\frac{\partial u_2}{\partial x^j}\,dx$$

for $u_1,u_2 \in H^1(\Omega)$.

Thus, for $u \in H^2(\Omega)$, $u = 0$ on $\partial\Omega$,

$$-\langle Au,u\rangle = B(u,u) + \sum_{i,j=1}^{n} \int_\Omega \left(\frac{\partial a_{ij}(x)}{\partial x^i}\right) u\frac{\partial u}{\partial x^j}\,dx + \sum_i \int_\Omega b_i u\frac{\partial u}{\partial x^i}\,dx + \int_\Omega cu^2\,dx.$$

Notice that since the a_{ij} are bounded,

$$|B(u_1,u_2)| \leqslant C\|u_1\|_{H^1}\|u_2\|_{H^1} ,$$

i.e., B is a continuous bilinear form on $H^1 \times H^1$.

We shall need two facts about elliptic operators. (Again, see the aforementioned references.)

1. (Gårding's Inequality) Let A be strongly elliptic. Then there are constants $c,d > 0$ such that

$$B(u,u) \geqslant c\|u\|_{H^1}^2 - d\|u\|_{L_2}^2, \quad u \in H^1(\Omega).$$

From this and the inequality $2ab \leqslant \varepsilon a^2 + \frac{1}{\varepsilon}b^2$, one deduces, equivalently,

$$-\langle Au,u\rangle \geqslant c_1\|u\|_{H^1}^2 - d_1\|u\|_{L_2}^2$$

for constants $c_1 > 0$, $d_1 > 0$ and $u \in H^2(\Omega)$, $u = 0$ on $\partial\Omega$.

2. (Lax-Milgram Theorem and Elliptic Estimates) If $\lambda > d_1$ and $f \in L_2(\Omega)$, there is a unique $u \in H^2(\Omega)$, $u = 0$ on $\partial\Omega$ such that

$$\lambda u - Au = f.$$

Notice that for $\lambda > d_1$, and $B_\lambda(u,v) = \langle \lambda u - Au, v\rangle$, $B_\lambda(u,u) \geqslant c_1 \|u\|^2_{H^1}$, so the Lax-Milgram theorem (see, e.g., Nirenberg [200]) can be applied to B_λ. The elliptic estimates then show that the resulting weak solution lies in $H^2(\Omega)$.

From these we deduce the following:

Theorem Let $X = L_2(\Omega)$ and A be as above. Then A generates a quasi-contractive semi-group† in X.

Proof. Clearly $D(A)$ is dense. Also, if $\lambda > d_1$,

$$\begin{aligned}\langle (\lambda - A)u, u\rangle &\geqslant c_1 \|u\|^2_{H^1} + (\lambda - d_1)\|u\|^2_{L_2} \\ &\geqslant (\lambda - d_1)\|u\|^2_{L_2},\end{aligned}$$

so we choose $\beta = d_1$. By property 2 above, $(\lambda - A)$ is surjective and hence by the second corollary of the Hille-Yosida theorem, $A \in G(X,1,\beta)$. As $(\lambda - A)$ is onto, A is automatically closed. □

The above example concerns the parabolic equation

$$\frac{\partial u}{\partial t} = Au.$$

We can also consider the hyperbolic equation

$$\frac{\partial^2 u}{\partial t^2} = Au,$$

using semigroup methods. Here we are concerned with the operator

$$A' = \begin{pmatrix} 0 & I \\ A & 0 \end{pmatrix}$$

---

†The semigroup is, in fact, analytic. However most hyperbolic type equations, such as those below, do not generate analytic semigroups (even if dissipation of the type occuring in elasticity is added).

with $D(A') = D(A) \times H^1(\Omega) \subset X = H^1(\Omega) \times L_2(\Omega)$.

Then $(\partial^2 u)/(\partial t^2) = Au$ is equivalent to

$$\frac{\partial}{\partial t}\begin{pmatrix} u \\ u_t \end{pmatrix} = A' \begin{pmatrix} u \\ u_t \end{pmatrix}.$$

<u>Theorem</u> With A as in the preceding theorem and with A' and X as just defined, A' generates a quasi-contractive semigroup on X.

<u>Proof.</u> By Gårding's inequality we can choose

$$B(u,u) + d\|u\|^2_{L_2} = |||u|||^2$$

to be an equivalent norm on $H^1(\Omega)$. Then using this norm:

$$\langle A'(u,\dot{u}),(u,\dot{u})\rangle = \langle(\dot{u},Au),(u,\dot{u})\rangle$$

$$= B(u,\dot{u}) + d\langle u,\dot{u}\rangle + \langle Au,\dot{u}\rangle$$

$$= d\langle u,\dot{u}\rangle - \sum_{i,j=1}^{n}\int_\Omega\left(\frac{\partial a_{ij}(x)}{\partial x^i}\right)u\frac{\partial u}{\partial x^j}\,dx$$

$$\leqslant C(|||u|||^2 + \|\dot{u}\|^2),$$

for a suitable constant C. The same estimate holds for $-A'$ since $\dot{u}$ can be replaced by $-\dot{u}$.

The solution of $(\lambda - A')(u,\dot{u}) = (f,\dot{f})$ is easily checked to be

$$u = (\lambda^2 - A)^{-1}\dot{f},$$

$$\dot{u} = f - \lambda u,$$

so $(\lambda - A')$ is onto for $|\lambda|$ sufficiently large. Thus, by the second corollary to the Hille-Yosida theorem, we have the result. □

The wave equation can also be dealt with using the following abstract theorem (Weiss [253], Goldstein [96]).

Theorem Let $\mathcal{H}$ be a (real) Hilbert space, B a self adjoint operator on $\mathcal{H}$ satisfying:

$$\langle Bx,x\rangle \geqslant c\langle x,x\rangle$$

for a constant $c > 0$. Let $B^{\frac{1}{2}}$ be a positive square root of B and let $\mathcal{H}_1$ be the domain of $B^{\frac{1}{2}}$ with the graph norm. Then the operator

$$A = \begin{pmatrix} 0 & I \\ -B & 0 \end{pmatrix}$$

generates a one parameter group on $\mathcal{H}_1 \times \mathcal{H}$ with domain $D(B)\times\mathcal{H}_1$.

The semigroup e^{tA} solves the abstract wave equation $(\partial^2 x)/(\partial t^2) = -Bx$. It is not hard to argue that A cannot be a generator on $\mathcal{H}\times\mathcal{H}$. (Indeed since $\begin{pmatrix} 0 & I \\ 0 & 0 \end{pmatrix}$ is bounded on $\mathcal{H}\times\mathcal{H}$, if A were a generator, so also would be $\begin{pmatrix} 0 & 0 \\ B & 0 \end{pmatrix}$. But this equation trivially integrates and one sees it is not a generator.)

Proof. Our condition on B means that the graph norm of $B^{\frac{1}{2}}$ is equivalent to the norm $|||x||| = \langle B^{\frac{1}{2}}x, B^{\frac{1}{2}}x\rangle$. Thus on $\mathcal{H}_1 \times \mathcal{H}$ we can take the Hilbert space norm

$$\|(x,y)\|^2 = \langle B^{\frac{1}{2}}x, B^{\frac{1}{2}}x\rangle + \langle y,y\rangle.$$

Provided that A is skew adjoint on $\mathcal{H}_1 \times \mathcal{H}$, the result would follow from the real form of Stone's theorem. Let us first check skew symmetry. Let $x_1, x_2 \in D_B$, $y_1, y_2 \in \mathcal{H}_1$. Then,

$$\begin{aligned}\langle A(x_1,x_2),(y_1,y_2)\rangle &= \langle (x_2,-Bx_1),(y_1,y_2)\rangle \\ &= \langle B^{\frac{1}{2}}x_2, B^{\frac{1}{2}}y_1\rangle + \langle -Bx_1, y_2\rangle\end{aligned}$$

$$= \langle Bx_2, y_1 \rangle - \langle Bx_1, y_2 \rangle,$$

since $x_2 \in D(B)$. Similarly we get

$$\langle (x_1,x_2), A(y_1,y_2) \rangle = \langle x_1, By_2 \rangle - \langle By_2, x_1 \rangle$$

so A is skew symmetric.

To show A is skew adjoint, let $(y_1,y_2) \in D(A^+)$ where A^+ denotes the skew adjoint of A. This means there is $(z_1,z_2) \in \mathcal{H}_1 \times \mathcal{H}$ such that

$$\langle A(x_1,x_2),(y_1,y_2) \rangle = - \langle (x_1,x_2),(z_1,z_2) \rangle$$

for all $(x_1,x_2) \in D(B) \times \mathcal{H}_1$. This assertion is equivalent to

$$\langle B^{\frac{1}{2}}x_2, B^{\frac{1}{2}}y_1 \rangle = - \langle x_2, z_2 \rangle \qquad \text{for all} \quad x_2 \in D(B^{\frac{1}{2}}),$$

and

$$\langle Bx_1, y_2 \rangle = \langle B^{\frac{1}{2}}x_1, B^{\frac{1}{2}}z_1 \rangle \qquad \text{for all} \quad x_1 \in D(B).$$

The first statement implies $B^{\frac{1}{2}}y_1 \in D(B^{\frac{1}{2}})$ or $y_1 \in D(B)$ and the second implies $y_2 \in D(B^{\frac{1}{2}})$. Hence $D(A^+) = D(A)$ so A is skew adjoint. □

The group generated by A can be written explicitly in terms of that generated by $C = B^{\frac{1}{2}}$ as

$$e^{tA} = \cosh(tC)\ (\text{Identity}) + \frac{\sinh(tC)}{C} A$$

where for example $\cosh tC = (e^{tC} + e^{-tC})/2$. Division by C is in terms of the operational calculus.

<u>Remarks</u> 1. The condition $C > 0$ can be relaxed if the spaces are modified as follows. Let B be self adjoint and non-negative, with trivial kernel and let $\mathcal{H}_B$ be the completion of $\mathcal{H}$ with respect to $\|x\|_B = \langle Bx, x \rangle$. Let $X = \mathcal{H}_B \times \mathcal{H}$ and let $A(x,y) = (y,-Bx)$. Then the closure of A is a

generator in X. The argument follows the lines above (see Weiss [253]). We shall prove an important converse to this result in Section 7.

2. The wave equation $u_{tt} = \Delta u$ on $\Omega \subset \mathbb{R}^n$ does not generate a semigroup in $W^{1,p} \times L^p$ if $p \neq 2$ and $n > 1$. See Littman [175].

We now study the symmetric hyperbolic systems of Friedrichs [93,94]. This type of system occurs in many problems of mathematical physics, e.g., Maxwell's equations; see Courant-Hilbert [58]. As we shall see in Section 7 this includes the equations of classical elasticity. As Friedrichs has shown, many nonlinear equations are also covered by systems of this type. For general relativity, see Fischer and Marsden [85]†. For the time-dependent and nonlinear cases, see Dunford and Schwartz [72] and Kato [137, 139]. We consider the equations in all of space for simplicity. We describe the general case in Section 7.

Let $u(x) \in \mathbb{R}^n$ for $x \in \mathbb{R}^m$ and consider the following initial value problem

$$a_0(x)\frac{\partial u}{\partial t} = \sum_{j=1}^{m} a_j(x)\frac{\partial u}{\partial x^j} + b(x)u(x) + f(x)$$

where a_0, a_j and b are $N \times N$ matrix functions. One assumes a_0 and a_j are symmetric and a_0 is uniformly positive definite, i.e., $a_0(x) \geqslant \varepsilon$ for some $\varepsilon > 0$. (This is a matrix inequality; it means $\langle a_0(x)\xi,\xi\rangle \geqslant \varepsilon\|\xi\|^2$ for all $\xi \in \mathbb{R}^N$.) In what follows we shall take $a_0 = \mathrm{id}$. The general case is dealt with in the same way by weighting the L_2 norm by a_0. We can assume $f = 0$ by our earlier remarks on inhomogeneous equations.

We make the following technical assumptions. The functions a_j, b are

---

†In all honesty we have to admit that we do not know how to put the equations of nonlinear elastodynamics into symmetric hyperbolic form.

to be of class C^1, uniformly bounded and with uniformly bounded first derivatives.

Theorem Under these assumptions, let $A_{min}: C_0^\infty \to L_2(\mathbb{R}^m,\mathbb{R}^N)$ (C_0^∞ denotes the C^∞ functions $u:\mathbb{R}^m \to \mathbb{R}^N$ with compact support) be defined by:

$$A_{min}u = \sum_{j=1}^{m} a_j(x)\frac{\partial u}{\partial x^j} + b(x)u(x).$$

Let A be the closure of A_{min}. Then A generates a quasi-contractive one parameter group in $L_2(\mathbb{R}^m,\mathbb{R}^N)$.

Proof. Define B_{min} on C_0^∞ by

$$B_{min}u = -\Sigma\frac{\partial}{\partial x^j}(a_j(x)u) + b(x)u.$$

Formally, B_{min} is the adjoint of A_{min} on C_0^∞. More precisely, it is easy to see that $A^*_{min} \supset B_{min}$. Let $A_{max} = B^*_{min}$ (In distribution language, A_{max} is just A_{min} defined on all u for which $A_{min}u$ lies in L_2, with derivatives in the sense of distributions).

We shall need the following.

Lemma A_{max} is the closure of A_{min}.

Proof. We shall sketch out the main steps. The method is often called that of the "Friedrichs Mollifier".

Let $u \in D_{A_{max}}$. We have to show there is $u_n \in C_0^\infty$ such that $u_n \to u$ and $A_{min}u_n \to A_{max}u \in L_2$.

Let $\rho:\mathbb{R}^m \to \mathbb{R}$ be C^∞ with support in the unit ball, $\rho \geqslant 0$ and

$\int \rho = 1$. Set $\rho_\varepsilon(x) = \frac{1}{\varepsilon^n}\rho(\frac{x}{\varepsilon})$ for $\varepsilon > 0$. Let $u_\varepsilon = \rho_\varepsilon * u$ (componentwise convolution).

We assert that $u_\varepsilon \longrightarrow u$ as $\varepsilon \longrightarrow 0$ (in L_2). Indeed, $\|u_\varepsilon\| \leqslant \|u\|$ so it is enough to check this for $u \in C_0^\infty$. Then it is a standard (and easy) argument; one obtains uniform convergence.

Now each u_ε is C^∞. Let L denote the differential operator

$$L = \Sigma a_j(x)\frac{\partial}{\partial x^j} + b(x).$$

Then one easily computes that

$$L(u_\varepsilon) = \int\{-\sum_j \frac{\partial}{\partial y^j}[a_j(y)\rho_\varepsilon(x-y)] + b(y)\rho_\varepsilon(x-y)\}u(y)dy$$

$$+ \int\{\Sigma \frac{\partial}{\partial y^j}([a_j(y) - a_j(x)]\rho_\varepsilon(x-y))$$

$$- [b(y) - b(x)]\rho_\varepsilon(x-y)\}u(y)dy.$$

The first term is just $\rho_\varepsilon * (A_{max}u)$ and thus we have proved $L(\rho_\varepsilon * u) - \rho_\varepsilon * A_{max}u \longrightarrow 0$ as $\varepsilon \longrightarrow 0$. It follows that $C^\infty \cap L_2$ is a <u>core</u> of A_{max}. That is, A_{max} restricted to $C_0^\infty \cap L_2 \cap D_{A_{max}}$ has closure A_{max}.

Let $\omega \in C_0^\infty(\mathbb{R}^m)$, ω with support in a ball of radius 2, $\omega \equiv 1$ on a ball of radius 1. Let $\omega_n(x) = \omega(x/n)$. Then $\omega_n u_\varepsilon \in C_0^\infty$ and

$$L(\omega_n u_\varepsilon) = \omega_n L u_\varepsilon + (\Sigma a_j(x)\frac{\partial \omega_n}{\partial x^j})u_\varepsilon.$$

As $n \longrightarrow \infty$ this converges to Lu_ε. This proves the lemma. □

Now we shall complete the proof. Let $A = A_{max}$. For $u \in C_0^\infty$, we get, as a_j is symmetric:

$$\langle Au,u\rangle = \int \sum_j \langle a_j \frac{\partial u}{\partial x^j}, u\rangle + \langle bu,u\rangle \, dx$$

$$= \int \{\Sigma \tfrac{1}{2}\frac{\partial}{\partial x^j}\langle a_j u,u\rangle - \tfrac{1}{2}\langle \frac{\partial a_j}{\partial x^j}u,u\rangle + \langle bu,u\rangle\} dx$$

$$= \int -\tfrac{1}{2}\langle \frac{\partial a_j}{\partial x^j}u,u\rangle + \langle bu,u\rangle \, dx.$$

Thus,

$$\langle Au,u\rangle \leqslant \beta_1 \int \langle u,u\rangle \, dx$$

where

$$\beta_1 = \sup\left(\tfrac{1}{2}\left|\frac{\partial a_j}{\partial x^j}\right| + |b|\right).$$

By the lemma this same inequality holds for all $u \in D(A)$. Thus,

$$\langle (\lambda - A)u,u\rangle \geqslant (\lambda - \beta_1)\langle u,u\rangle$$

from which it follows that

$$\|(\lambda - A)u\| \geqslant (\lambda - \beta_1)\|u\|.$$

Thus $(A + \lambda)$ has closed range if $\lambda > \beta_1$, and is one to one. To show the range is the whole space we must show that

$$(\lambda - A)^*\omega = 0 \quad \text{implies} \quad \omega = 0.$$

($(\lambda - A)^*\omega = 0$ means ω is orthogonal to the range.) But B = closure of B_{min} (defined in the proof of the lemma) equals A^*. Thus $(\lambda - B)\omega = 0$. As above, we have

$$\|(\lambda - B)\omega\| \geqslant (\lambda - \beta_2)\|\omega\|$$

so $(\lambda - B)\omega = 0$ implies $\omega = 0$ for $\lambda > \beta_2$. For $\beta = \sup(\beta_1, \beta_2)$ then $\lambda > \beta$ implies

$$\|(\lambda - A)^{-1}\| \leq 1/(\lambda - \beta).$$

Since the conditions on A are unaffected by replacing A with $-A$ we see that

$$\|(A + \lambda)^{-1}\| \leq 1/(|\lambda| - \beta), \qquad |\lambda| > \beta.$$

Hence $-A$ generates a quasi-contractive group. □

Provided the coefficients are smooth enough, it is not hard to argue that A generates a semi group on H^s as well as on $H^0 = L_2$. (H^s is the Hilbert space of functions whose derivatives of order $\leq s$ are in L_2.) This follows by showing the H^s norm remains bounded under the flow on L_2 (Use Gronwall's inequality).

Let us return to the wave equation to see how it can be covered by this theorem. Consider the system:

$$\begin{cases} \dfrac{\partial v^0}{\partial t} = v^{n+1}, \\[2ex] \dfrac{\partial v^1}{\partial t} = \dfrac{\partial v^{n+1}}{\partial x^1}, \\ \qquad\vdots \\ \dfrac{\partial v^n}{\partial t} = \dfrac{\partial v^{n+1}}{\partial x^n}, \\[2ex] \dfrac{\partial v^{n+1}}{\partial t} = \dfrac{\partial v^1}{\partial x^1} + \dots + \dfrac{\partial v^n}{\partial x^n}. \end{cases}$$

Here $a_1 = \begin{pmatrix} 0 & 0 & \dots & 0 \\ 0 & 0 & \dots & 1 \\ 0 & 0 & \dots & 0 \\ \vdots & & & \\ 0 & 1 & \dots & 0 \end{pmatrix}$ etc., so our system is symmetric. By the preceding theorem, it generates a group. Let $u, \dot{u} \in H_1 \times L_2$ and consider the initial data

$$v^0 = u, \quad v^1 = \frac{\partial u}{\partial x^1}, \; \dots \; v^n = \frac{\partial u}{\partial x^n}, \quad v^{n+1} = \dot{u}.$$

Then the equations for v reduce exactly to the wave equation for u, so $(\partial^2 u)/(\partial t^2) = \Delta u$ generates a group on $H^1 \times L_2$ as before.

The final example in this section is a fifth order equation which occurs in supersonic flow over a vibrating panel (see Dowell [71]).

<u>Example</u> Consider small vibrations of a panel, as shown in Fig. 9. Neglecting nonlinear and two dimensional effects (see Holmes and Marsden [128] for a more general case), the equations are:

$$\ddot{v} + \alpha \dot{v}'''' + v'''' - \Gamma v'' + \rho v' + \sqrt{\rho\delta}\, \dot{v} = 0,$$

where $v(t,x)$ is the panel deflection, $\dot{} = \partial/\partial t$ and $' = \partial/\partial x$. Here α is a viscoelastic structural damping constant, ρ is an aerodynamic pressure, Γ is an in-plane tensile load and $\sqrt{\rho\delta}$ is aerodynamic damping. We assume $\alpha > 0$, $\delta > 0$, $\rho > 0$.

If the edges of the plate are simply supported, we impose the boundary conditions $v = 0$, $v'' + \alpha \dot{v}'' = 0$ at $x = 0,1$.

We choose

$$H_0^2 = \{u \in H^2([0,1]) \mid u = 0 \quad \text{at} \quad x = 0,1\},$$

and let $X = H_0^2 \times L_2$.

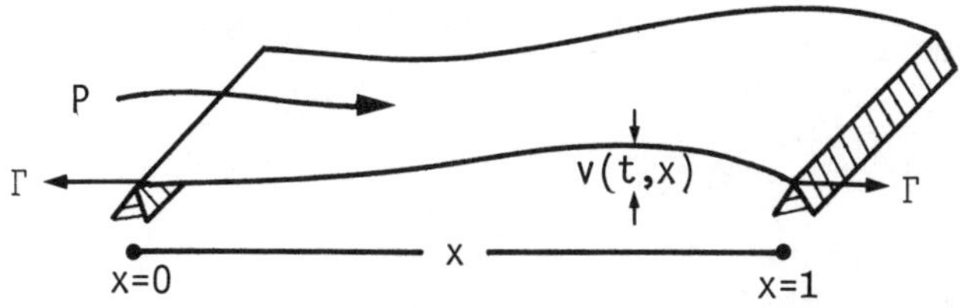

Fig. 9.

Let

$$A\begin{pmatrix} v \\ \dot{v} \end{pmatrix} = \begin{pmatrix} \dot{v} \\ -\alpha\dot{v}'''' - v'''' \end{pmatrix}$$

with $D(A) = \{(v,\dot{v}) \in H_0^2 \times L_2 \mid v + \alpha\dot{v} \in H^4,\ \ v'' + \alpha\dot{v}'' = 0$ at $x = 0,1,$ $\dot{v} \in H_0^2\}$. On X we choose the inner product

$$\langle (v,\dot{v}),(w,\dot{w}) \rangle = (v'',w'') + (\dot{v},\dot{w})$$

where $(\ ,\ )$ denotes the L_2 inner product.

We assert that A generates a contraction semigroup in X. Indeed,

$$\left\langle A\begin{pmatrix} v \\ \dot{v} \end{pmatrix}, \begin{pmatrix} v \\ \dot{v} \end{pmatrix} \right\rangle = \left\langle \begin{matrix} \dot{v} \\ -\alpha\dot{v}'''' - v'''' \end{matrix}, \begin{matrix} v \\ \dot{v} \end{matrix} \right\rangle$$

$$= (\dot{v}'',v'') - (\alpha\dot{v}'''' + v'''', \dot{v})$$

$$= (\dot{v}'',v'') - (\alpha\dot{v}'' + v'', \dot{v}'')$$

$$= -\alpha(\dot{v}'',\dot{v}'') \leqslant 0.$$

Next we show that $(\lambda - A)$ is onto for $\lambda > 0$. This follows the same pattern as the previous examples.

First of all, the range is closed; let $x_n = (v_n, \dot{v}_n) \in D(A)$ and suppose $y_n = (\lambda - A)(v_n, \dot{v}_n) \longrightarrow y \in X$. From the above estimate, and Schwarz's inequality,

$$\|\lambda x_n - Ax_n\|_X \geqslant \lambda \|x_n\|,$$

from which it follows that x_n converges to say x in X. Since $y_n \longrightarrow y$, Ax_n converges as well. Thus $\alpha\dot{v}_n + v_n$ converges in H^4, so $x \in D(A)$ and $Ax = y$.

Secondly, the range of A is dense. Suppose that there is a $y \in X$ such that

$$\langle (\lambda - A)x, y \rangle = 0 \qquad \text{for all } x \in D(A).$$

Thus if $x = (v, \dot{v})$, $y = (w, \dot{w})$, then

$$(\lambda v'' - \dot{v}'', w'') = 0$$

and

$$(\lambda \dot{v} + \alpha \dot{v}'''' + v'''', \dot{w}) = 0.$$

Set $\dot{v} = 0$; the first equation gives $w'' = 0$, so $w = 0$. The second equation with $\dot{v} = 0$ shows $\dot{w}$ satisfies $\dot{w}'''' = 0$ in the weak sense, so is smooth. Setting $v = 0$ and $\dot{v} = \dot{w}$ then shows $\dot{w} = 0$ since $\lambda > 0$, $\alpha > 0$.

Actually, the origin is globally attracting for A. That is, in a suitable equivalent norm

$$\|e^{tA}\|_X \leqslant e^{-\varepsilon t}, \qquad \varepsilon > 0.$$

This is because the spectrum of A (computed by separation of variables) is discrete and consists of eigenvalues

$$\lambda_j = -\frac{j^4\pi^4\alpha}{2}\{1 \pm \sqrt{1 - \frac{4}{\alpha^2\pi^4 j^4}}\}, \qquad j = 1,2,\ldots$$

with $\mathrm{Re}\lambda_j \leq \max\{-\frac{\alpha\pi^4}{2}, -\frac{1}{\alpha}\}$. (See Remark 6, Trend to Equilibrium above.)

For the general equations, observe that

$$B\begin{pmatrix} v \\ \dot{v} \end{pmatrix} = \begin{pmatrix} 0 \\ \Gamma v'' + \rho v' + \sqrt{\rho}\delta\dot{v} \end{pmatrix}$$

is a bounded operator in X and therefore $A+B$ generates a C^0 semigroup. For $\rho \neq 0$, it is not simple to compute the spectrum of $A+B$ explicitly; however, one can do so numerically and determine thereby for which Γ, ρ, δ, α the origin is attracting. The estimates (in the spectral norm)

$$\|e^{t(A+B)}\|_X \leq e^{(\|B\| - \varepsilon)t}$$

and $\|v'\|^2_{L_2} \leq \|v''\|^2_{L_2}/\pi^2$ give parts of this region. (The term $\sqrt{\rho}\delta\dot{v}$ can be omitted from B for this as it is dissipative.) We refer to Holmes and Marsden [128] and Parks [208] for further details.

6 LINEAR HAMILTONIAN SYSTEMS

At this point we present the theory of linear Hamiltonian systems using the semigroup theory developed in the preceding section. The nonlinear theory will be given in Section 9. We shall show that classical linear elasticity is a Hamiltonian system, but shall postpone a detailed discussion of the existence and uniqueness theory for linear elastodynamics until the next section, where the equations will also be coupled to dissipative mechanisms.

The abstract theory of Hamiltonian systems presented here is useful for seeing how elasticity fits into the general theory, and for best seeing how Hamiltonian techniques may be applied. Even in linear elasticity the ideas can be useful. We point out an example following the definition of a linear Hamiltonian system.

<u>Definitions</u> Let X be a Banach space. A linear (weak) <u>symplectic structure</u> on X is a continuous bilinear map

$$\Omega : X \times X \longrightarrow \mathbb{R}$$

which is

(i) skew; i.e., $\Omega(u,v) = -\Omega(v,u)$

and (ii) weakly non-degenerate; i.e., $\Omega(u,v) = 0$ for all $v \in X$ implies $u = 0$.

We often speak of X with a symplectic structure Ω as a <u>phase space</u>.

A linear operator $A : D(A) \longrightarrow X$ is called <u>Hamiltonian</u> if it is Ω-skew, i.e.,

$$\Omega(Ax,y) = -\Omega(x,Ay)$$

for all $x,y \in D(A)$.

The <u>Hamiltonian or energy function</u> of A is defined by

$$H(u) = \tfrac{1}{2}\Omega(Au,u), \qquad u \in D(A).$$

A bounded linear map $T \in B(X)$ is called a <u>canonical transformation</u> if it preserves Ω, i.e.,

$$\Omega(Tu,Tv) = \Omega(u,v)$$

for all $u,v \in X$.†

Here is an indication of how Hamiltonian ideas can be of use in under-

---

†In the pull-back notation of Section 1, this reads $T^*\Omega = \Omega$.

standing elasticity (in addition to the Hamiltonian nature of the dynamics). Suppose there is a linear relationship amongst the phase variables defining a linear subspace $L \subset X$. This relationship is called reciprocal if Ω vanishes on $L \times L$ and L is maximal with respect to the property (in finite dimensions this means $\dim L = \frac{1}{2} \dim X$).[†]

Let us suppose, throughout this section that A is a closed operator. Thus $D(A)$ becomes a Banach space if we use the graph norm:

$$|||u|||^2 = \|Au\|^2 + \|u\|^2.$$

Clearly, $H : D(A) \longrightarrow \mathbb{R}$ is a differentiable function and

$$dH(u)\cdot v = \Omega(Au, v)$$

which uniquely determines A on its domain. Conversely, we have the following:

Proposition Let A be a closed operator and $H : D(A) \longrightarrow \mathbb{R}$ a C^1 function such that

$$dH(u)\cdot v = \Omega(Au, v)$$

for all $u, v \in D(A)$. Then A is Hamiltonian with energy H, after modifying H by a constant.

Proof. From the above formula H is C^2 and

$$d^2H(0)(u, v) = \Omega(Au, v).$$

[†]See R. Abraham and J. Marsden [1] for further details. For examples in linear elasticity, see Gurtin [106], p.66,98,101,207,218 and in linear thermoelasticity, see Carlson [34], p.320.

Since this is symmetric in (u,v), A is Ω-skew, so is Hamiltonian. Since $dH(u)\cdot v = d[\frac{1}{2}\Omega(Au,u)]\cdot v$, H is the energy, up to a constant. □

Examples 1. (a) Let X_1, Y_1 be Banach spaces with Y_1 continuously and densely included in X_1. Let $X = Y_1 \times X_1^*$ and set $\Omega : X \times X \longrightarrow \mathbb{R}$,

$$\Omega((y,\alpha),(\overline{y},\overline{\alpha})) = \overline{\alpha}(y) - \alpha(\overline{y}).$$

Then Ω is a weak symplectic form.†

(b) If $\langle\ ,\ \rangle$ is a weakly non-degenerate symmetric bilinear form on X_1 and $X = Y_1 \times X_1$, then

$$\Omega((y,x),(\overline{y},\overline{x})) = \langle \overline{x},y\rangle - \langle x,\overline{y}\rangle$$

is a weak symplectic form.

In example (a) we call Ω the canonical symplectic form induced on X and in (b) we say it is the symplectic form induced by the metric.

In example (a) note that we can write

$$\Omega((y,\alpha),(\overline{y},\overline{\alpha})) = -(\overline{y},\overline{\alpha})\cdot J\cdot\binom{y}{\alpha},$$

where $J = \begin{pmatrix} 0 & I \\ -I & 0 \end{pmatrix}$. In this case A is formally related to H by $A(y,\alpha) = J\cdot dH(y,\alpha)$, as is readily checked.

(c) If $\mathcal{H}_1$ is real Hilbert space and we let $\mathcal{H} = \mathcal{H}_1 \oplus \mathcal{H}_1$ be its complexification then the canonical symplectic form induced on $\mathcal{H}$ in (b) (with $Y_1 = \mathcal{H}$, $X_1 = \mathcal{H}$) can be written

$$\Omega : \mathcal{H}\times\mathcal{H} \longrightarrow \mathbb{R}, \quad \Omega(\phi,\psi) = -\mathrm{Im}\langle\phi,\psi\rangle,$$

---

†If $Y_1 = X_1$, Ω is non-degenerate in the strong sense; i.e., the induced map of X to X^* is an isomorphism if and only if X_1 is reflexive; i.e., the canonical inclusion $X_1 \subset X_1^{**}$ is onto.

as is easily checked.

2. Let $X = \mathbb{R}^n \times (\mathbb{R}^n)^*$ with g^{ij} and c_{ij} symmetric matrices. Let points in X be denoted by $(x,p) = (x^i, p_j)$ and set

$$H(x^i, p_j) = \tfrac{1}{2} g^{ij} p_i p_j + \tfrac{1}{2} c_{ij} x^i x^j,$$

and let

$$\Omega((x,p),(\overline{x},\overline{p})) = \overline{p}_i x^i - p_i \overline{x}^i.$$

The corresponding linear Hamiltonian operator is

$$A(x,p) = \left(\frac{\partial H}{\partial p_i}, -\frac{\partial H}{\partial x_i}\right)$$

$$= (g^{ij} p_j, -c_{ij} x^j).$$

If P is an $n \times n$ matrix diagonalizing $C = (c_{ij})$, i.e., if $P^{-1}CP = \Lambda = (\lambda_i)$ is diagonal, then

$$T : X \longrightarrow X$$

$$T(x,p) = (P^{-1}x, p \cdot P^T)$$

is a canonical transformation uncoupling the terms $c_{ij} x^i x^j$. If g^{ij} is positive definite we can simultaneously diagonalise g^{ij} and c_{ij} and thereby uncouple the equations into n harmonic oscillators. (See Abraham and Marsden [1, Section 5.6] for further details.)

3. The wave equation $\ddot{\phi} = \Delta\phi$ is Hamiltonian on $X = H^1(\mathbb{R}^n) \times L^2(\mathbb{R}^n)$ with

$$H(\phi, \dot{\phi}) = \tfrac{1}{2}\int_{\mathbb{R}^n} |\dot{\phi}|^2 dx + \tfrac{1}{2}\int_{\mathbb{R}^n} |\nabla\phi|^2 dx,$$

$$\Omega((\phi,\dot\phi),(\psi,\dot\psi)) = \int (\langle \dot\psi,\phi\rangle - \langle \dot\phi,\psi\rangle)dx$$

(the symplectic form of 1(b) with $Y_1 = H^1$, $X_1 = L_2$ and $\langle\ ,\ \rangle$ the L_2 inner product) and

$$A(\phi,\dot\phi) = \begin{pmatrix} \dot\phi \\ \Delta\phi \end{pmatrix},$$

with $D(A) = H^2(\mathbb{R}^n) \times H^1(\mathbb{R}^n)$.

4. If H_{op} is a symmetric operator on complex Hilbert space $\mathcal{H}$, the abstract Schrodinger equation,

$$\frac{\partial\psi}{\partial t} = iH_{op}\psi$$

is Hamiltonian with

$$H(\psi) = -\tfrac{1}{2}\langle H_{op}\psi,\psi\rangle,$$

$$\Omega(\psi,\phi) = \operatorname{Im}\langle\psi,\phi\rangle, \qquad \text{(See example 1(c))},$$

$$A\psi = iH_{op}\psi.$$

If H_{op} is self-adjoint, then Stone's theorem guarantees that A generates a one parameter unitary group, as explained in Section 5.

Now we turn to a few general relationships between Hamiltonian generators and the corresponding one parameter groups (or semigroups).

Proposition Suppose $A \in G(X,M,\beta)$ and is Hamiltonian with energy H. Then if $U(t)$ is the semigroup generated by A,

(i) each $U(t)$ is a canonical transformation,

(ii) $H(U(t)x) = H(x)$ (conservation of energy).

Proof. (i) If $u,v \in D(A)$, then

$$\frac{d}{dt}\Omega(U(t)u,\ U(t)v) = \Omega(AU(t)u,\ U(t)v) + \Omega(U(t)u,\ AU(t)v) = 0,$$

since A is Ω-skew. Hence,

$$\Omega(U(t)u,\ U(t)v) = \Omega(u,v)$$

for $u,v \in D(A)$ and hence everywhere by continuity and denseness of $D(A)$.

(ii)
$$\begin{aligned} H(U(t)u) &= \tfrac{1}{2}\Omega(AU(t)u,\ U(t)u) \\ &= \tfrac{1}{2}\Omega(U(t)Au,\ U(t)u) \\ &= \tfrac{1}{2}\Omega(Au,u) \\ &= H(u). \quad \square \end{aligned}$$

One half of Stone's theorem asserts that the generator of a one parameter group of isometries on a Hilbert space is skew-adjoint. The following is a generalization of that fact.

Proposition Let $U(t)$ be a one parameter group of canonical transformations on (X,Ω) with generator A. Then A is Ω-skew adjoint.

Proof. (Nelson) Let $A^\dagger$ be the Ω-adjoint of A. Since $\Omega(U(t)x,\ U(t)y) = \Omega(x,y)$, we have

$$\Omega(Ax,y) + \Omega(x,Ay) = 0, \qquad x,y \in D(A),$$

so $A^\dagger \supset -A$. Now let $f \in D(A^\dagger)$, $A^\dagger f = g$. For $x \in D(A)$, write

$$U(t)x = x + \int_0^t AU(s)x\,ds,$$

so

$$\Omega(U(t)x,f) = \Omega(x,f) + \int_0^t \Omega(AU(s)x,f)ds.$$

Thus

$$\Omega(x,U(-t)f) = \Omega(x,f) + \int_0^t \Omega(x,U(-s)A^\dagger f)ds.$$

Since $D(A)$ is dense,

$$U(-t)f = f + \int_0^t U(-s)A^\dagger f\,ds.$$

It follows that $f \in D(A)$ and $-Af = A^\dagger f$. □

The equation $\ddot{\phi} = -\Delta\phi$ shows that A being Ω-skew adjoint is not sufficient for A to be a generator. However it becomes sufficient if we impose a positivity condition. This result is due to Chernoff-Marsden [39].

Theorem Let X be a Banach space and Ω a weak symplectic form on X. Let A be an Ω-skew adjoint operator in X and set

$$[x,y] = \Omega(Ax,y),$$

the energy inner product. Assume

$$[x,x] \geqslant c\|x\|_X \qquad \text{(stability)}$$

for a constant $c > 0$.

Let $\mathcal{H}$ be the completion of $D(A)$ with respect to $[\ ,\ ]$ and let

$$D(\tilde{A}) = \{x \in D(A) \mid Ax \in \mathcal{H}\}$$

and

$$\tilde{A}x = Ax, \quad x \in D(\tilde{A}).$$

Then $\tilde{A}$ generates in $\mathcal{H}$ a one parameter group of canonical transformations (relative to $\tilde{\Omega}$, the restriction of Ω to $\mathcal{H}$) and these are, moreover, isometries (relative to the energy inner product on $\mathcal{H}$).

<u>Proof</u>. Because the energy inner product satisfies $[x,x] \geq c\|x\|_X^2$, we can identify $\mathcal{H}$ with a subspace of X. Relative to $[\ ,\ ]$ we note that $\tilde{A}$ is skew-symmetric: for $x,y \in D(\tilde{A})$,

$$[x,\tilde{A}y] = \Omega(Ax,Ay) = -\Omega(Ay,Ax) = -[y,\tilde{A}x] = -[Ax,y].$$

We next shall show that $\tilde{A}$ is skew-adjoint. To do this, it is enough to show that $\tilde{A} : D(\tilde{A}) \longrightarrow \mathcal{H}$ is onto. This will follow if we can show that $A : D(A) \longrightarrow X$ is onto.

Let $w \in X$. By the Riesz theorem, there is an $x \in \mathcal{H}$ such that

$$\Omega(w,y) = [x,y], \quad \text{for all} \quad y \in \mathcal{H}.$$

In particular,

$$\Omega(w,y) = \Omega(Ay,x) = -\Omega(x,Ay) \quad \text{for all} \quad y \in D(A).$$

Therefore $x \in D(A^\dagger) = D(A)$ and $Ax = w$. Thus A is onto.

It remains to show that $\tilde{\Omega}_1$ is left invariant by $U(t) = e^{t\tilde{A}}$. For this we need only verify that $\tilde{A}$ is $\tilde{\Omega}$-skew: for $x,y \in D(\tilde{A})$,

$$\tilde{\Omega}(\tilde{A}x,y) = \Omega(Ax,y) = -\Omega(x,Ay) = -\tilde{\Omega}(x,\tilde{A}y). \quad \square$$

<u>Example</u> (<u>Abstract Wave Equation</u>) Let $\mathcal{H}$ be a real Hilbert space and B a self adjoint operator satisfying $B \geq C > 0$. Then

$$A = \begin{pmatrix} 0 & I \\ -B & 0 \end{pmatrix}$$

is Hamiltonian on

$$X = D(B^{\frac{1}{2}}) \times \mathcal{H}$$

with

$$D(A) = D(B) \times D(B^{\frac{1}{2}}),$$

$$\Omega((x_1,y_1),(x_2,y_2)) = \langle y_2,x_1\rangle - \langle x_2,y_1\rangle$$

and energy

$$H(x,y) = \tfrac{1}{2}\|y\|^2 + \tfrac{1}{2}\langle Bx,x\rangle .$$

The above theorem reproduces the theorem we proved in the preceding section on the abstract wave equation $\ddot{x} = -Bx$. It follows from the above theorem that the corresponding one parameter group consists of canonical transformations which preserve energy.

Example (Classical Elastodynamics) We now show how the (homogeneous) equations of classical linear elastodynamics, viz.,

$$\rho\ddot{u} = \operatorname{div}(\mathbb{c}\cdot\nabla u),$$

i.e.,

$$\rho\ddot{u}^a = (c^{abcd}\, u_{c|d})_{|b},$$

are a Hamiltonian system. We work in a region Ω with displacement or traction boundary conditions imposed, and assume, as in Section 4, that c^{abcd} comes from a stored energy function; i.e., $c^{abcd} = c^{badc}$.

We choose $X = H^1 \times L_2$,

$$\Omega((u,\dot{u}),(v,\dot{v})) = \int_\Omega \rho(\dot{v}u - \dot{u}v)dx,$$

$$A(u,\dot{u}) = \begin{pmatrix} \dot{u} \\ \frac{1}{\rho}\operatorname{div}(\mathbb{c}\cdot\nabla u)\end{pmatrix},$$

and

$$H(u,\dot{u}) = \frac{1}{2}\int_\Omega \rho|\dot{u}|^2\,dx + \frac{1}{2}\int_\Omega c^{abcd}e_{ab}e_{cd}\,dx,$$

where $e_{ab} = \frac{1}{2}(u_{a|b} + u_{b|a})$, as in Section 4.

Now A has the form $\begin{pmatrix} 0 & I \\ -B & 0 \end{pmatrix}$ where $Bu = \frac{1}{\rho}\operatorname{div}(\mathbf{c}\cdot\nabla u)$ is a symmetric operator in L_2 with inner product

$$\langle u_1,u_2\rangle = \int_\Omega \rho(x)u_1(x)u_2(x)\,dx.$$

Thus we have a Hamiltonian system. It will generate a one parameter group if we can show the energy is positive definite (and the remaining technical conditions hold).

This leads naturally to the condition that there is an $\varepsilon > 0$ such that for any <u>symmetric</u> tensor e_{ab},

$$c^{abcd}e_{ab}e_{cd} \geqslant \varepsilon\|e\|^2 \qquad \text{(stability condition)}^\dagger$$

This condition implies (but is not implied by) the <u>strong ellipticity condition</u>

$$c^{abcd}\xi_a\xi_c\eta_b\eta_d \geqslant \varepsilon\|\xi\|^2\|\eta\|^2$$

(take $e_{ab} = \xi_a\eta_b + \xi_b\eta_a$) and thus allows the existence and uniqueness theory developed for scalar equations in the previous section to go through. Alternatively, it allows the hypotheses in the abstract theorem developed above to be verified. Details are given in the following section.

For isotropic elasticity, $c_{abcd} = \lambda g_{ab}g_{cd} + 2\mu g_{ac}g_{bd}$ as we saw in Section 4. Here,

---

† That this implies positive definiteness of the energy for the displacement problem relies on Korn's inequality. See Sections 7 and 8.

$$c^{abcd}e_{ab}e_{cd} = \lambda(\mathrm{tr}\, e)^2 + 2\mu e^2$$

and a simple exercise in completing the square shows that the stability condition holds if and only if $(3\lambda + 2\mu)/3 > 0$ and $\mu > 0$. However, as we can guess from our work in Section 5 and as we shall see in Section 7, generation of a semigroup only requires the weaker strong ellipticity condition, which for the isotropic case becomes, as is easily checked (see Gurtin [106], p.85-87):

$$\lambda + 2\mu > 0, \quad \mu > 0.$$

In Sections 9 and 10 we shall derive conservation laws for general nonlinear elasticity. They apply, in particular, to linear elasticity. In preparation for that it seems appropriate to make a few comments here on conservation laws in the framework of abstract linear Hamiltonian systems.

We let X be a Banach space with symplectic form Ω and let A and B be two Hamiltonian operators in X with corresponding energy functions

$$H_A(u) = \tfrac{1}{2}\Omega(Au,u), \quad H_B(u) = \tfrac{1}{2}\Omega(Bu,u)$$

as above. The Poisson bracket is defined by

$$\{H_A,H_B\}^\circ(u) = \Omega(Au,Bu), \quad u \in D(A) \cap D(B).$$

(The reader can check that for finite dimensional systems this is the usual expression

$$\{f,g\} = \sum_{i=1}^{n}\left(\frac{\partial f}{\partial q^i}\frac{\partial g}{\partial p_i} - \frac{\partial f}{\partial p_i}\frac{\partial g}{\partial q^i}\right).$$

See Abraham and Marsden [1] for details.)

Proposition If $x \in D([A,B])$, we have

$$\{H_A, H_B\} = H_{[A,B]}$$

where $[A,B] = AB - BA$ is the operator commutator.

Proof. By definition,

$$\begin{aligned}\{H_A, H_B\}(x) &= \Omega(Ax, Bx) \\ &= \tfrac{1}{2}\Omega(Ax, Bx) - \tfrac{1}{2}\Omega(Bx, Ax) \\ &= -\tfrac{1}{2}\Omega(BAx, x) + \tfrac{1}{2}\Omega(ABx, x) \\ &= \tfrac{1}{2}\Omega([A,B]x, x) \\ &= H_{[A,B]}(x). \quad \square\end{aligned}$$

Now suppose that A generates a group or semigroup U_t and B generates a group V_t. As we have seen these are necessarily symplectic transformations.

Theorem Suppose that V_t is a symmetry group of the function H_A in the following sense: each map V_t leaves $D(A)$ invariant, and $H_A \circ V_t = H_A$. Then H_B is a constant of the motion; that is, U_t leaves $D(B)$ invariant and $H_B \circ U_t = H_B$. Moreover, U and V commute; that is, $U_s V_t = V_t U_s$ for all s,t.

Proof. Fix t. Then for each $x \in D(A)$ we have

$$\Omega(Ax, x) = \Omega(AV_t x, V_t x) = \Omega(V_{-t} A V_t x, x).$$

From the polarization identity it follows that

$$A = V_{-t} A V_t.$$

But $V_{-t}AV_t$ generates the group (or semigroup)

$$s \longrightarrow V_{-t}U_sV_t.$$

Accordingly we must have $U_s = V_{-t}U_sV_t$. That is, $U_sV_t = V_tU_s$. From this it follows that $U_sD(B) \subset D(B)$. Finally, we have the relation, for $x \in D(B)$,

$$U_sBx = \frac{d}{dt} U_sV_tx \,|\,_{t=0} = \frac{d}{ds} V_tU_sx \,|\,_{t=0} = BU_sx.$$

Hence

$$H_B(U_sx) = \tfrac{1}{2}\Omega(BU_sx,U_sx) = \tfrac{1}{2}\Omega(U_sBx,U_sx)$$

$$= \tfrac{1}{2}\Omega(Bx,x) = H_B(x). \quad \square$$

In order to conclude that the flows U_s and V_t commute, it is <u>not</u> enough to have $\{H_A,H_B\} = 0$, i.e., $[A,B] = 0$. In fact Nelson [198] has given a counter-example: two skew-adjoint operators A,B such that $[A,B] \equiv 0$ on $D(AB) \cap D(BA)$, but such that e^{sA} and e^{tB} do <u>not</u> commute. Thus the infinite dimensional case is much subtler than the finite dimensional case, and it is well to be wary of reliance on formal calculations alone.

As we indicated in the earlier example dealing with the Schrodinger equation, there are usually complex structures lurking in the background for real linear Hamiltonian systems. In fact, under fairly general circumstances one can show that a real linear Hamiltonian System can be given a complex structure in which the group is unitary so its generator must be self adjoint (See Cook [57] and Chernoff-Marsden [39], Sect. 2.7). Actually in the context of second order systems, this idea in a version due to Weiss [252] will be useful for elastodynamics. It is related to the problem of showing that Cauchy elasticity cannot give rise to a bounded group. Details are given in Section 7.

In the previous section we saw that the wave equation can be written as a symmetric hyperbolic system. The same is true of classical elastodynamics. This gives an alternative approach to the existence and uniqueness problem.

<u>Example</u> (<u>Classical Elastodynamics as a Symmetric Hyperbolic System</u>); The basic equations in a region $\Omega \subset \mathbb{R}^3$, viz.,

$$\rho \frac{\partial^2 u_i}{\partial t^2} = \frac{\partial}{\partial x_j}\left(c_{ijkl} \frac{\partial u_k}{\partial x_l}\right) + \rho b_i \qquad \text{(summation implied)}$$

can be written in symmetric hyperbolic form as follows (c.f., Brockway [30] and Wilcox [256]).

Let $u = (u,T,v)$ where $u = (u_1,u_2,u_3)$, $T = (T_{11},T_{22},T_{33},T_{12},T_{13},T_{23})$ and $v = (v_1,v_2,v_3)$. Let $e = (e_{11},e_{22},e_{33},\gamma_{12},\gamma_{13},\gamma_{23})$ where $\gamma_{ij} = 2e_{ij}$. The constitutive equation $T_{ij} = c_{ijkl}e_{kl}$ can be put in the matrix form $T = \mathbb{c}e$. We will need the condition that $\mathbb{c}$ is symmetric and uniformly positive definite, so $\mathbb{c}$ is invertible. As above, these requirements are $c_{ijkl} = c_{klij}$ and there is an $\varepsilon > 0$ such that $c_{ijkl}(x)e_{ij}e_{kl} \geqslant \varepsilon\, e_{ij}e_{ij}$ for all symmetric e_{ij} and all $x \in \Omega$. Define

$$a_0 = \begin{pmatrix} I & \cdot & \cdot \\ \cdot & \mathbb{c}^{-1} & \cdot \\ \cdot & \cdot & I \end{pmatrix}, \quad a_i \frac{\partial}{\partial x^i} = \begin{pmatrix} \cdot & \cdot & \cdot \\ \cdot & \cdot & D \\ \cdot & D^T & \cdot \end{pmatrix},$$

where I is the 3×3 identity matrix a dot denotes "zero", and

$$D = \begin{pmatrix} \partial/\partial x_1 & \cdot & \cdot \\ \cdot & \partial/\partial x_2 & \cdot \\ \cdot & \cdot & \partial/\partial x_3 \\ \partial/\partial x_2 & \partial/\partial x_1 & \cdot \\ \partial/\partial x_3 & \cdot & \partial/\partial x_1 \\ \cdot & \partial/\partial x_3 & \partial/\partial x_2 \end{pmatrix}.$$

Let

$$b = \begin{pmatrix} \cdot & \cdot & I \\ \cdot & \cdot & \cdot \\ \cdot & \cdot & \cdot \end{pmatrix}, \quad \text{and} \quad f = \begin{pmatrix} \cdot \\ \cdot \\ \rho b \end{pmatrix},$$

where $b = (b_1, b_2, b_3)$.

It is readily verified that the basic equations now have the symmetric hyperbolic form

$$a_0 \frac{\partial u}{\partial t} = a_i \frac{\partial u}{\partial x_i} + bu + f.$$

Thus for $u \in L_2(\mathbb{R}^3)$ we get a one parameter group, by the general theory for symmetric hyperbolic systems given in Section 5.

For the full initial boundary value problem, we refer to Hughes and Marsden [131] where sharp regularity results are obtained. However, it is considerably more technical than the second order approach, because the latter can make use of the deep results on elliptic boundary value problems. The second also has the advantage that it only requires <u>strong ellipticity</u> and not <u>stability</u>.

We saw above that the equations of elasticity form a Hamiltonian system. For conditions under which symmetric hyperbolic systems in general are Hamiltonian, see Chernoff and Marsden [39]. See also Chernoff [38] for an intrinsic treatment of symmetric hyperbolic systems on manifolds.

7 EXISTENCE AND UNIQUENESS FOR LINEAR ELASTODYNAMICS

The plan for the first part of this section is summarized in the table below. The section begins by giving further credence to the idea that Cauchy elasticity is unphysical. Using a result of Weiss [253] we show that Cauchy elasticity can never generate a contractive semigroup unless it is actually hyperelasticity. In the dynamical sense, a contraction semigroup means dynamical stability. Thus Cauchy elasticity cannot yield stable dynamics. Following this we examine the case of hyperelasticity.

A main result in hyperelasticity is that the strong ellipticity condition is necessary and sufficient for the generation of a contractive semigroup on $X = H^1 \times L_2$. This is proved by combining results from elliptic theory with the second corollary to the Hille-Yosida theorem in Section 5.

The fact that the equations generate a semigroup in X embodies the idea that we have a continuous linear dynamical system in X. In particular the solutions depend continuously in X as the initial conditions are varied in the same space X. This is to be compared with other types of continuous data dependence where the solution and initial data vary in different spaces. For the latter, strong ellipticity is not required. See Knops and Payne [151] (and related references in the bibliography) for extensive discussion of these points.

We show that stability in the sense that energy is positive definite relative to the H^1 topology is equivalent to dynamical stability; i.e., a contractive semigroup. The use of Korn's inequalities to prove stability for classical elasticity from positive definiteness of the elasticity tensor c_{ijkl} (acting on pairs of symmetric tensors) is discussed.

As we saw in the previous section, stability is a stronger assumption than strong ellipticity. When a bifurcation occurs in the nonlinear theory, one expects the linearized theory to lose stability, but not strong ellipticity. When stability is lost, directions of exponential growth (the $e^{\beta t}$ growth of the semigroup) will develop. This is perfectly consistent with strong ellipticity.

If strong ellipticity "strictly" fails, we will show that the equations cannot generate a semigroup on <u>any</u> space $Y \times L_2$, where $Y \subset L_2$. This result was suggested by N.S. Wilkes [258-260] using a logarithmic convexity argument due to Knops and Payne [149]. The overall situation is summarized as follows:

<u>Linear Hyperelasticity</u>

strong ellipticity	$\Longleftrightarrow$	the equations of motion generate a quasi-contractive semigroup in $H^1 \times L_2$ (relative to some Hilbert space structure).
$\Uparrow$		$\Uparrow$
stability (classical elasticity)	$\Longleftrightarrow$	generates a <u>contractive</u> semigroup on $H^1 \times L_2$ (relative to some Hilbert space norm).
$\Uparrow$ (Korn)		
elasticity tensor positive definite		

- - - - - - - - - - - - - - -

strong ellipticity "strictly" fails	$\Longleftrightarrow$	no semigroup on any space $Y \times L_2$, $Y \subset L_2$.

- - - - - - - - - - - - - - -

<u>Cauchy Elasticity</u>

generation of a contractive semigroup on a Hilbert space $Y \times L_2$, $Y \subset L_2$	$\Longrightarrow$	hyperelasticity.

Following these results we give a few abstract theorems relevant to hyperelastic systems with dissipation. These are inspired by Dafermos [61] and are complemented by an appendix by C. Navarro.

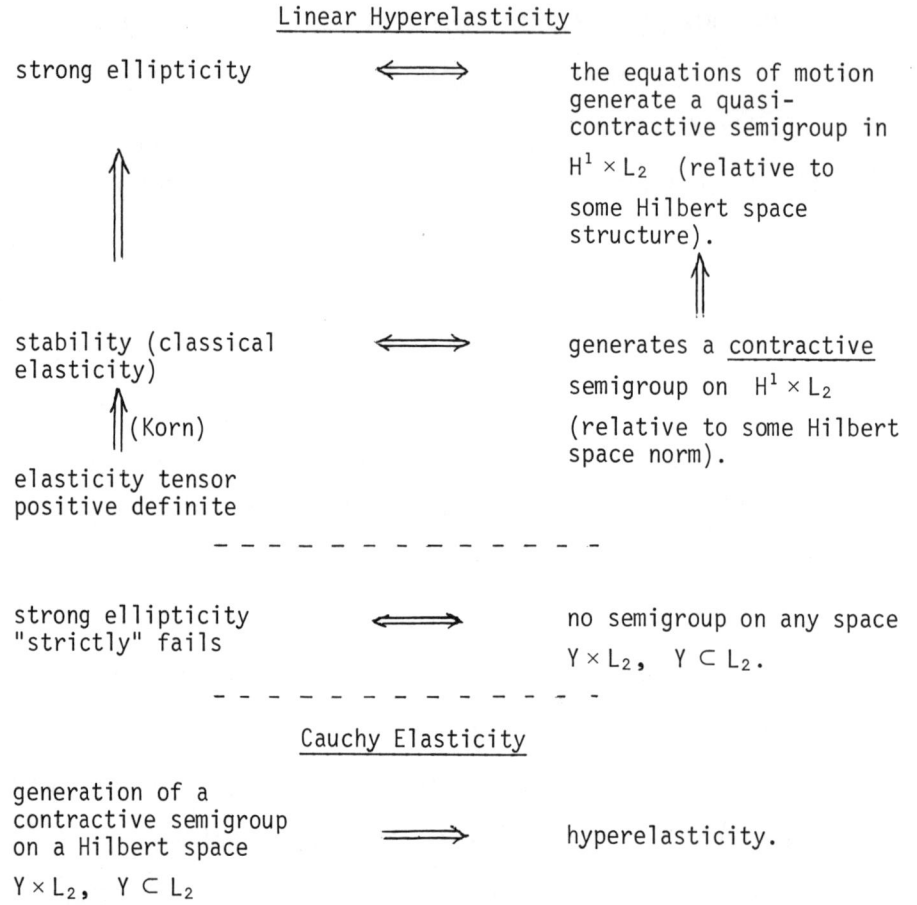

Many of the results presented here for existence and uniqueness of linear elastodynamics are well-known (see, Fichera [83]), although their proof has not usually employed the semigroup approach. The results here on the necessity of strong ellipticity and hyperelasticity seem to be new. They give further weight to the value of the semigroup approach.

Consider then the equations of linearized elasticity on a bounded smooth region† $\Omega \subset \mathbb{R}^n$

$$\rho\ddot{u} = \operatorname{div}(a \cdot \nabla u),$$

i.e.,

$$\rho\ddot{u}_i = (a_{ijkl}u_{k,l})_{,j} \qquad \text{(summation implied)},$$

and work in Euclidean coordinates for simplicity. There is no essential difficulty in working covariantly. We have dropped the inhomogeneous terms $f = -\rho\mathring{a} + \rho b + \operatorname{div}(\mathring{t})$ for they cause no added difficulty in questions of existence and uniqueness, as we saw in Section 5.

The boundary conditions are assumed to be either

$$\text{displacement:} \quad u = 0 \quad \text{on } \partial\Omega,$$

$$\text{traction:} \quad a \cdot \nabla u = 0 \quad \text{on } \partial\Omega,$$

again taken to be homogeneous without loss of generality. We assume a_{ijkl} are C^2 and $\rho(x) \geqslant \delta > 0$ is C^0.

We recall that the material in question is <u>hyperelastic</u> when $a_{ijkl} = a_{jilk}$. This is easily seen to be equivalent to <u>symmetry</u> of the operator

$$Au = \frac{1}{\rho}\operatorname{div}(a \cdot \nabla u),$$

---

†For $\Omega = \mathbb{R}^n$ things simplify slightly. For general unbounded regions the theory presumably goes through, using the relevant results from elliptic theory (for instance, those of Browder [31] or Agmon, Douglis and Nirenberg [3]).

in L_2 with norm $\langle u,v\rangle = \int_\Omega \rho uv dx$.

We rewrite, as usual, the equations of motion as $\ddot{u} = Au$ or

$$\frac{d}{dt}\begin{pmatrix} u \\ \dot{u} \end{pmatrix} = A' \begin{pmatrix} u \\ \dot{u} \end{pmatrix} \quad \text{where} \quad A' = \begin{pmatrix} 0 & I \\ A & 0 \end{pmatrix}.$$

The domain of A is taken to be $H^2(\Omega)$ with the appropriate boundary conditions imposed.

Let us begin by disposing of Cauchy elasticity. This can be done by the following abstract result for Hamiltonian systems.

Theorem (Weiss [253]). Let A be a linear operator in Hilbert space $\mathcal{H}$ with domain $D(A)$. Let Y be a Hilbert space,

$$D(A) \subset Y \subset \mathcal{H}.$$

Assume

$$A' = \begin{pmatrix} 0 & I \\ A & 0 \end{pmatrix}, \quad D(A') = D(A) \times Y$$

generates a contractive semigroup on $X = Y \times \mathcal{H}$. Then A is a self-adjoint operator, and in particular is symmetric.

Proof. By the Lumer-Phillips theorem of Section 5, we have

$$\langle (u,\dot{u}), A'(u,\dot{u})\rangle_X \leqslant 0,$$

i.e., $\langle u,\dot{u}\rangle_Y + \langle \dot{u}, Au\rangle_{\mathcal{H}} \leqslant 0$.

Since this holds for all $u \in D(A)$, $\dot{u} \in Y$, we can replace $\dot{u}$ by $-\dot{u}$. The left side changes sign, so we must have

$$\langle u,\dot{u}\rangle_Y + \langle \dot{u}, Au\rangle_{\mathcal{H}} = 0.$$

Thus,

$$\langle \dot{u}, Au \rangle_{\mathcal{H}} = - \langle u, \dot{u} \rangle_Y,$$

and so A is symmetric and non-positive. It is also self-adjoint since $(\lambda - A)$ is surjective for $\lambda > 0$. □

This result shows that linear Cauchy elasticity can never lead to a stable dynamical system in $Y \times L_2$, unless it is hyperelastic. This is presumably an undesirable situation for Cauchy elasticity. It is a semigroup analogue of the usual work theorems which are used to cast doubt on Cauchy elasticity (cf. Gurtin [106], p.82). Therefore, from now on we will assume we are dealing with hyperelasticity.

The above theorem has another interesting corollary. It shows that we are forced to choose the Y norm to be the energy norm: for the contractive case

$$\|u\|_Y^2 = - \langle u, Au \rangle_{\mathcal{H}},$$

and that our semigroups are forced to be groups of isometries.

Remarks 1. Weiss [253] also shows that one is forced into working on Hilbert space as opposed to general Banach spaces.

2. Related to the contractive assumption is an abstract result of Nagy [196], namely that a bounded one parameter group on Hilbert space is actually unitary in an equivalent Hilbert norm.

We say that a_{ijkl} is strongly elliptic† when there is an $\varepsilon > 0$ such that

---

†The strong ellipticity condition is closely related to the reality of wave speeds; see Gurtin [106], §§70,71.

$$a_{ijkl}(x)\xi_i\xi_k\eta_j\eta_l \geqslant \varepsilon|\xi|^2\,|\eta|^2; \qquad \xi,\eta \in \mathbb{R}^n, \quad x \in \bar{\Omega}.$$

We say that strong ellipticity <u>strictly fails</u> if there exist ξ and η and a point x for which

$$a_{ijkl}(x)\xi_i\xi_k\eta_j\eta_l < 0.$$

Let $H^1(\Omega)$ from now on, stand for $H_0^1(\Omega)$ (i.e., with $u=0$ on $\partial\Omega$) in the displacement case and just $H^1(\Omega)$ in the traction case. Thus $H^1(\Omega)$ is the H^1 completion of $D(A)$.

<u>Theorem</u> The operator $A' = \begin{pmatrix} 0 & I \\ A & 0 \end{pmatrix}$ on $X = H^1(\Omega) \times L_2(\Omega)$ with domain $D(A) \times H^1(\Omega)$ generates a quasi-contractive group (relative to some inner product on X) if and only if a_{ijkl} is strongly elliptic.

To prove this we need two results from elliptic theory, as we did for scalar equations in Section 5.

<u>Gårding's Inequality</u> a_{ijkl} is strongly elliptic if and only if there are constants $c > 0$ and $d > 0$ such that

$$B(u,u) \geqslant C\|u\|^2_{H^1} - d\|u\|^2_{L_2}$$

for all $u_i \in H^1(\Omega)$, where

$$B(u,v) = \int_\Omega a_{ijkl}u_{k,l}v_{i,j}\,dx.$$

<u>Lax-Milgram and Elliptic Estimates</u> If a_{ijkl} is strongly elliptic, $\lambda > d$ and $f \in L_2(\Omega)$ there is a unique $u \in H^2(\Omega)$ satisfying the boundary conditions such that

$$\lambda u - Au = f.$$

It should be noted that these results are for systems and as such are more delicate than the scalar versions. The difficult parts are Gårding's inequality and the elliptic estimates. Their proofs for general elliptic boundary value problems (systems) may be found in Morrey [191]. (The elliptic estimates are also found in Agmon, Douglis and Nirenberg [3]).

Proof of Theorem. First assume a_{ijkl} are strongly elliptic. Notice that for $v \in D(A)$, $u \in H^1(\Omega)$,

$$B(u,v) = -\langle u, Av\rangle$$

where the L_2 inner product is weighted with ρ, as above. By Gårding's inequality,

$$B(u,u) + d\|u\|^2_{L_2} = |||u|||^2$$

is equivalent to the H^1 norm. We use it on $H^1 \times L_2$. Then

$$\begin{aligned}\langle A'(u,\dot u),(u,\dot u)\rangle &= \langle \dot u, Au\rangle_{L_2} + \langle u,\dot u\rangle_{H^1}\\ &= B(u,\dot u) + d\langle u,\dot u\rangle_{L_2} + \langle Au,\dot u\rangle_{L_2}\\ &= d\langle u,\dot u\rangle_{L_2}\\ &\leqslant \tfrac{1}{2}d(\|u\|^2 + \|\dot u\|^2)\\ &\leqslant \tfrac{1}{2}d(|||u|||^2 + \|\dot u\|^2).\end{aligned}$$

The same estimate holds for $-A'$ since $\dot u$ can be replaced by $-\dot u$.

We know that $\lambda u - Au = f$ is solvable for u if $\lambda > d$, and the solution of

$$(\lambda - A')(u,\dot u) = (f,\mathring{f})$$

is

$$u = (\lambda^2 - A)^{-1} f,$$

$$\dot{u} = f - \lambda u,$$

so $(\lambda - A')$ is onto for $|\lambda|$ sufficiently large. Thus by the second corollary of the Hille-Yosida theorem, A' is the generator of a quasi-contractive semigroup.

Next we prove the converse. The Lumer-Phillips theorem implies that

$$\langle u,\dot{u}\rangle_{H^1} + \langle Au,\dot{u}\rangle_{L_2} \leqslant \beta\{\langle u,u\rangle_{H^1} + \langle \dot{u},\dot{u}\rangle_{L^2}\}$$

for all $u \in D(A)$, $\dot{u} \in H^1$. Letting $\dot{u} = \alpha u$, $\alpha > 0$, gives

$$-\langle Au,u\rangle \geqslant \frac{1}{\alpha}\{\alpha\langle u,u\rangle_{H^1} - \beta\langle u,u\rangle_{H^1} - \beta\alpha^2\langle u,u\rangle_{L_2}\}.$$

Choosing $\alpha > \beta$ we see that

$$-\langle Au,u\rangle \geqslant c\|u\|^2_{H^1} - d\|u\|^2_{L^2},$$

where $c = (\alpha-\beta)/\alpha$, $d = \beta\alpha$. Thus Gårding's inequality holds and so we have strong ellipticity (The latter result is due, essentially, to Hadamard)[†]. □

We note that, as remarked above, generation of a semigroup <u>automatically</u> implies generation of a group for second order systems under consideration.

It should be noted that this result <u>automatically</u> gives us a sharp regularity result, at least if the elliptic estimates are assumed; i.e., if $u \in H^2(\Omega)$ and u satisfies the boundary conditions, then for $s \geqslant 2$,

$$\|u\|_{H^s} \leqslant C(\|Au\|_{H^{s-2}} + \|u\|_{L_2}).$$

[†]The <u>argument</u> in Truesdell and Noll [248], p.253 will give the result stated here.

This shows that $D(A^m) \subset H^{2m}$(which, by the Sobolev embedding theorem, lies in C^1 if $1 < 2m - n/2$). From the abstract theory of semigroups in Section 5 we know that if $(u(0),\dot{u}(0)) \in D(A')^m$, then $(u(t),\dot{u}(t)) \in D(A')^m$ as well. For instance $(u(0),\dot{u}(0)) \in D(A')^3$ means

$$u(0) \in D(A^2) \quad \text{and} \quad \dot{u}(0) \in D(A).$$

Note that this automatically means $u(0)$ and $\dot{u}(0)$ must satisfy extra boundary conditions; in general, these extra conditions for $(u,\dot{u})$ to belong to $D(A')^m$ are called the <u>compatibility conditions</u>.

In particular, if $u(0)$ and $\dot{u}(0)$ are C^∞ in x and belong to the domain of every power of A, then the solutions are C^∞ in (x,t) in the classical sense.

For later use, we make an observation about second order systems in general. Namely, if A' generates a quasi-contractive semigroup on $Y \times \mathcal{H}$, then necessarily, A generates one on $\mathcal{H}$.

If we have

$$a_{ijkl}(x)\xi_i\xi_k\eta_j\eta_l \geqslant \varepsilon(x)\,|\xi|^2\,|\eta|^2,$$

and $\varepsilon(x)$ vanishes at some points, then one can still, under technical conditions sufficient to guarantee A is selfadjoint (see, e.g., Reed and Simon [217] and references therein), get a quasi-contractive semi-group on $Y \times L_2$, where Y is the completion of $H^1(\Omega)$ in the energy norm. One can argue, along the lines above, that if A' generates a quasi-contractive semi-group on $Y \times L_2$, then we must have

$$a_{ijkl}(x)\xi_i\xi_k\eta_j\eta_l \geqslant 0.$$

By the remark just made, without assuming $Y = H^1(\Omega)$, one cannot hope to use well-posedness to prove strong ellipticity.

Next we wish to study the relationship between strong ellipticity and stability. We first define stability in terms of positive definiteness of the energy and then show that this is equivalent to dynamical stability.[†]

We say that a_{ijkl} is <u>stable</u> provided that there is a $c > 0$ such that

$$B(u,u) \geqslant C\|u\|^2_{H^1} \quad \text{for all} \quad u \in H^1(\Omega),$$

i.e., the elastic potential energy is positive definite relative to the H^1 norm.

<u>Proposition</u> A' generates a contractive group on $H^1 \times L_2$ (relative to some inner product on H^1) if and only if we have stability.

<u>Proof</u>. In the above proof of sufficiency we saw we could take $\beta = \frac{d}{2}$. But stability is precisely the condition $d = 0$, so we get a contractive group. Note that since stability implies Gårding's inequality, we have strong ellipticity automatically.

Conversely, if we get a contractive group relative to some equivalent inner product $\langle\langle\ ,\ \rangle\rangle_{H^1}$ on H^1, we saw in Weiss' theorem that we must have

$$B(u,u) = \langle\langle u,u \rangle\rangle_{H^1}$$

which implies stability. □

For the displacement problem in classical elasticity, <u>positive definiteness of the elasticity tensor</u>, i.e., there is a $\delta > 0$ such that

$$c_{ijkl}e_{ij}e_{kl} \geqslant \delta \|e\|^2$$

[†]This is usually called the <u>energy criterion</u> for stability. It is discussed in Section 11 for nonlinear systems.

for all symmetric e_{ij} implies stability[†] by virtue of Korn's first inequality

$$\int_\Omega e^2 dx \geqslant c\|u\|^2_{H^1}, \qquad c > 0,$$

where $e_{ij} = \frac{1}{2}(u_{i,j} + u_{j,i})$ and $u = 0$ on $\partial\Omega$. By considering isotropic classical elasticity, strong ellipticity does not imply stability. Korn's first inequality is readily proved using Fourier series and a partition of unity argument for reduction to the case of a torus.

For the traction problem, Korn's first inequality cannot hold since e_{ij} is invariant under the Euclidean group. Instead one has Korn's second inequality (see Fichera [83], Friedrichs [92], etc.)

$$\int_\Omega e^2 dx + \int_\Omega u^2 dx \geqslant c\|u\|^2_{H^1}.$$

As it stands, this shows that positive definiteness of the elasticity tensor implies Gårding's inequality. However, we already know Gårding's inequality is true from strong ellipticity alone. Thus Korn's second inequality does not seem to be needed. Nevertheless, there is a deeper reason for Korn's second inequality which is not usually appreciated. Namely, if we view the traction problem as a Hamiltonian system and move into centre of mass and constant angular momentum "coordinates",[††] then in the appropriate quotient space of $H^1 \times L_2$, we get a new Hamiltonian system and in this quotient space, Korn's second inequality can be interpreted as saying that positive definiteness of the elasticity tensor implies stability. Since this is a

---

[†] We saw in Section 6 that strong ellipticity is implied as well.

[††] In Hamiltonian systems language this is a special case of the general procedure of "reduction". See Abraham and Marsden [1] and Marsden-Weinstein [186].

rather involved, but straightforward story, details are omitted. (Consult Gurtin [106], p.104)

Next we sketch out an argument due to Wilkes [258-260] based on logarithmic convexity ideas (see Knops and Payne [49]) to show:

<u>Theorem</u> If the strong ellipticity condition strictly fails then A' cannot generate a semi-group on $Y \times L_2$ where $D(A) \subset Y \subset L_2$.

<u>Proof</u>. When strong ellipticity strictly fails, as defined above, the argument used to prove Hadamard's theorem shows that

$$\inf_{\|u\|_{L_2}=1} -\langle u, Au \rangle = -\infty$$

(Roughly speaking, one can rescale $u(x)$ keeping its L_2 norm constant, but blowing up its H^1 norm.) Suppose A' generates a semigroup $U(t)$ of type (M,β). Then we can choose $u(0)$, $\dot{u}(0)$ such that

$$2\langle \dot{u}(0), u(0) \rangle > \beta, \ \|u(0)\|^2_{L_2} = 1,$$

and

$$\tfrac{1}{2}\langle \dot{u}(0), \dot{u}(0) \rangle - \tfrac{1}{2}\langle u(0), Au(0) \rangle = c < 0.$$

Let $(u(t),\dot{u}(t)) = U(t)(u(0),\dot{u}(0))$ and $F(t) = \frac{1}{2}\langle u(t), u(t) \rangle$. Then clearly

$$\dot{F} = \langle \dot{u}, u \rangle \quad \text{and} \quad \ddot{F} = \langle u, Au \rangle + \langle \dot{u}, \dot{u} \rangle.$$

Note that c is the initial energy, and the energy is constant in time. Then by Schwarz's inequality,

$$\frac{\dot{F}^2}{F} = \frac{2\langle \dot{u}, u \rangle^2}{\langle u, u \rangle} \leqslant 2\|\dot{u}\|^2$$

$$= 2c + \ddot{F}.$$

Thus,

$$F\ddot{F} - \dot{F}^2 \geqslant -4cF \geqslant 0.$$

Hence,

$$\frac{d^2}{dt^2}(\log F) \geqslant 0,$$

and so,

$$F(t) \geqslant F(0)\exp\left(\frac{\dot{F}(0)}{F(0)}t\right),$$

i.e.,

$$\|u\|^2_{L_2} \geqslant \|u(0)\|^2_{L_2}\, e^{\gamma t}$$

where $\gamma = 2\langle \dot{u}(0), u(0)\rangle / \langle u(0), u)0)\rangle$. But because $U(t)$ is a semigroup of type (M,β), the Y topology is stronger than the L_2 topology and $\gamma > \beta$, and such an inequality is impossible. □

This concludes our discussion of the Cauchy problem for linear elastic systems. Next we consider a couple of abstract examples of dissipative mechanisms which can be added to the conservative equations. These results are inspired by those of Weiss [252] and Dafermos [61]. In the appendix to this section the ideas are applied to linear thermoelastic materials with memory.

First we consider dissipation of rate type; i.e., the equation

$$\ddot{u} = Au + B\dot{u}.$$

Theorem Suppose A and B generate (quasi)-contractive semigroups on a Hilbert space $\mathcal{H}$ and $D(B) \subset D(A)$. Suppose $A' = \begin{pmatrix} 0 & I \\ A & 0 \end{pmatrix}$ generates a (quasi)-contractive semigroup on $Y \times \mathcal{H}$ with domain $D(A) \times Y$, $D(A) \subset Y$. Let $C = \begin{pmatrix} 0 & I \\ A & B \end{pmatrix}$ with $D(C) = D(A) \times D(B)$. Then $\overline{C}$ generates a (quasi)-contractive semigroup on $Y \times \mathcal{H}$.

Proof. We need the preliminary result:

Lemma (Trotter [244]). Let A and B generate (quasi)-contractive semigroups on a Banach space X and let $D(B) \subset D(A)$. Then there is a $\delta > 0$ such that $cA + B$ generates a quasi-contractive semigroup if $0 \leq c \leq \delta$.

We refer to Trotter's paper for the proof.

Proof of Theorem. Let $\beta > 0$ be such that

$$\langle (u,\dot{u}), A'(u,\dot{u}) \rangle_{Y\times\mathcal{H}} \leq \beta \|(u,\dot{u})\|^2_{Y\times\mathcal{H}}$$

and $\gamma > 0$ be such that

$$\langle B\dot{u},\dot{u}\rangle_{\mathcal{H}} \leq \gamma \|\dot{u}\|^2_{\mathcal{H}}.$$

Then if $B' = \begin{pmatrix} 0 & 0 \\ 0 & B \end{pmatrix}$,

$$\langle (u,\dot{u}),(A'+B')(u,\dot{u})\rangle_{Y\times\mathcal{H}} \leq \beta(\|u\|^2_Y + \|\dot{u}\|^2_{\mathcal{H}}) + \gamma\|\dot{u}\|^2_{\mathcal{H}} \leq \rho\|(u,\dot{u})\|^2_{Y\times\mathcal{H}},$$

where $\rho = \beta + \gamma$. By the second corollary to the Hille-Yosida theorem, it remains to show that $\lambda - C = \lambda - A' - B'$ has dense range if λ is sufficiently large. Suppose $(v,\dot{v})$ is orthogonal to the range of $\lambda - C$. Then

$$\langle \lambda u - \dot{u}, v\rangle_Y + \langle \lambda\dot{u} - Au - B\dot{u},\dot{v}\rangle_{\mathcal{H}} = 0.$$

Setting $\dot{u} = \lambda u$, we get

$$\langle \lambda^2 u - Au - \lambda Bu, \dot{v}\rangle = 0,$$

i.e.,

$$\langle \lambda u - \frac{1}{\lambda}Au - Bu, \dot{v}\rangle = 0.$$

If $\lambda > \delta^{-1}$, where δ is given in the lemma, we conclude that $\lambda - A/\lambda - B$ is onto, so $\dot{v} = 0$. Returning to the original orthogonality condition, we get $v = 0$. □

Note that if A is symmetric, so that $\ddot{u} = Au$ is Hamiltonian and $B \leqslant 0$, then the energy is decreasing:

$$\frac{1}{2}\frac{d}{dt}(\langle \dot{u},\dot{u}\rangle - \langle u,Au\rangle) = \langle \dot{u},B\dot{u}\rangle \leqslant 0.$$

This is the usual situation for rate type dissipation.

For example, we can conclude that if a_{ijkl} is strongly elliptic, then

$$\rho\ddot{u}_i = (a_{ijkl}u_{k,l})_{,i} + \dot{u}_{i,jj},$$

i.e.,

$$\rho\ddot{u} = \mathrm{div}(a\cdot\nabla u) + \Delta\dot{u}$$

with say displacement boundary conditions, generates a quasi-contractive semigroup on $X = H^1 \times L_2$. If the elastic energy is positive definite, i.e., stability holds, then the semigroup is contractive (it does not seem to be analytic). One can establish trend to equilibrium results either by spectral methods (see Section 5) or by Liapunov techniques (see Dafermos [61]). The theorem can also be applied to the panel equation in Section 5.

If the dissipation is of thermal type, the equations take the form

$$\ddot{u} = Au + B\theta,$$
$$\dot{\theta} = C\theta + D\dot{u}.$$

We make these assumptions:

(i) $A' = \begin{pmatrix} 0 & I \\ A & 0 \end{pmatrix}$ generates a quasi-contractive semigroup on $Y \times \mathcal{H}$.

(ii) C is a non-positive self adjoint operator on a Hilbert space $\mathcal{H}_\theta$.

(iii) B is an operator from $\mathcal{H}_\theta$ to $\mathcal{H}$ and is densely defined. Moreover, $D = -B^*$ and is densely defined.

(iv) $D(A) \subset D(D) \subset Y$.

(v) $D(C) \subset D(B)$.

(vi) $B(1-C)^{-1}D$, a non-positive symmetric operator, has self-adjoint closure (i.e., is essentially selfadjoint). [In the example below $B(1-C)^{-1}D$ is bounded].

Let $G = \begin{pmatrix} 0 & I & 0 \\ A & 0 & B \\ 0 & D & C \end{pmatrix}$ with domain

$$D(A) \times D(D) \times D(C) \subset Y \times \mathcal{H} \times \mathcal{H}_\theta = X,$$

so that $\frac{d}{dt}\begin{pmatrix} u \\ \dot{u} \\ \theta \end{pmatrix} = G\begin{pmatrix} u \\ \dot{u} \\ \theta \end{pmatrix}$ represents the above system.

<u>Theorem</u> Under these assumptions, $\overline{G}$ generates a quasi-contractive semigroup on X. (If A' generates a contractive semigroup, so does $\overline{G}$).

<u>Proof</u>. We have, in the inner product on $Y \times \mathcal{H} \times \mathcal{H}_\theta$,

$$\begin{aligned}
\langle (u,\dot{u},\theta), G(u,\dot{u},\theta) \rangle &= \langle (u,\dot{u},\theta), (\dot{u}, Au + B\theta, C\theta + D\dot{u}) \rangle \\
&= \langle u,\dot{u} \rangle_Y + \langle \dot{u}, Au + B\theta \rangle_{\mathcal{H}} + \langle \theta, C\theta + D\dot{u} \rangle_{\mathcal{H}_\theta} \\
&\leqslant \beta \| (u,\dot{u}) \|^2_{Y \times \mathcal{H}} + \langle \dot{u}, B\theta \rangle + \langle \theta, D\dot{u} \rangle + \langle \theta, C\theta \rangle \\
&= \beta \| (u,\dot{u}) \|^2_{Y \times \mathcal{H}} + \langle \theta, C\theta \rangle \\
&\leqslant \beta \| (u,\dot{u}) \|^2_{Y \times \mathcal{H}} \leqslant \beta \| (u,\dot{u},\theta) \|^2_X ,
\end{aligned}$$

so $(G - \beta)$ is dissipative. By the second corollary to the Hille-Yosida theorem, it remains to show that for λ sufficiently large, $(\lambda - G)$ has dense range. Let $(v,\dot{v},g)$ be orthogonal to the range:

$$\langle \lambda u - \dot{u}, v \rangle + \langle \lambda\dot{u} - Au - B\theta, \dot{v} \rangle + \langle \lambda\theta - C\theta - D\dot{u}, g \rangle = 0.$$

For $u \in D(A)$, let $\dot{u} = \lambda u$ and $\theta = \lambda(\lambda - C)^{-1}Du$. Then

$$\langle \lambda^2 u - Au - \lambda B(\lambda - C)^{-1}Du, \dot{v} \rangle = 0.$$

By using Trotter's result and the same argument as in the preceding theorem, if λ is sufficiently large,

$$\lambda^2 - A - \lambda B(\lambda - C)^{-1}D$$

will have dense range. Thus $\dot{v} = 0$. Taking $\dot{u}$ and $\theta = 0$, one sees that $v = 0$ and taking $u = 0 = \dot{u}$, one gets $g = 0$. □

For example, if a_{ijkl} is strongly elliptic we find that

$$\rho\ddot{u} = \operatorname{div}(a \cdot \nabla u) + m\nabla\theta,$$

$$c\dot{\theta} = k\Delta\theta + \frac{m}{\rho}\nabla\cdot\dot{u}, \quad u,\theta = 0 \text{ on } \partial\Omega,$$

generates a quasi-contractive semigroup on $H^1 \times L_2 \times L_2$ (with the L_2 spaces appropriately weighted), where $c,k > 0$, $m > 0$.[†]

Finally we consider visco-elasticity of memory (or Boltzman) type. The equations now have the form

$$\ddot{u} = Au + Bw,$$

$$\dot{w} = Cw.$$

We make these assumptions

(i) $A' = \begin{pmatrix} 0 & I \\ A & 0 \end{pmatrix}$ generates a quasi-contractive semigroup on $Y \times \mathcal{H}$.

(ii) C generates a contractive semigroup on $\tilde{\mathcal{H}}$.

[†]Above we saw that "well-posedness" of the elastic part implied a is strongly elliptic. The Clausius-Duhem inequality implies $k \geqslant 0$. Well-posedness of the heat part then implies $c > 0$ and then of the whole implies $m \geqslant 0$.

(iii) There is an injection $i : \mathcal{H} \longrightarrow \tilde{\mathcal{H}}$ (constant histories) such that $C \circ i = 0$.

(iv) B is a densely defined operator of $\tilde{\mathcal{H}}$ to $\mathcal{H}$ such that $i \circ B$ is symmetric and non-negative, $D(B) \subset D(C)$, $D(B)$ is a core for C and B is one to one.

(v) Let $\mathfrak{a}(w_1,w_2) = \langle i \cdot Bw_1, w_2 \rangle_{\tilde{\mathcal{H}}}$ for $w_1,w_2 \in D(B)$. Suppose that on $X = Y \times \mathcal{H} \times \tilde{\mathcal{H}}$,

$$\| (u,\dot{u},w) \|_X^2 = \| (u,\dot{u} \|_{Y \times \mathcal{H}}^2 + \mathfrak{a}(iu - w, iu - w) - \mathfrak{a}(iu,iu)$$

is an inner product equivalent to the original one. Let

$$G = \begin{pmatrix} 0 & I & 0 \\ A & 0 & B \\ 0 & 0 & C \end{pmatrix}$$

which is the operator on X corresponding to the above equations, with domain $D(A) \times Y \times D(B)$.

(vi) $\langle iBw, Cw \rangle \geqslant 0$ for all $w \in D(B)$.

Theorem Under these assumptions, $\overline{G}$ generates a quasi-contractive semi-group on X (contractive if A' generates a contractive semigroup).

Proof. We have, in the X inner product of (v),

$$\begin{aligned} \langle (u,\dot{u},w),(\dot{u},Au + Bw,Cw) \rangle &= \langle u,\dot{u} \rangle + \langle \dot{u},Au \rangle + \langle \dot{u},Bw \rangle \\ &\quad + \langle iB(iu - w), i\dot{u} - Cw \rangle - \langle iBiu,iu \rangle \\ &= \langle u,\dot{u} \rangle + \langle \dot{u},Au \rangle + \langle iB(iu - w), -Cw \rangle \\ &= \langle u,\dot{u} \rangle + \langle \dot{u},Au \rangle - \langle iB(iu - w), C(iu - w) \rangle \end{aligned}$$

$$\leqslant \langle u,\dot{u}\rangle + \langle \dot{u},Au\rangle$$

$$\leqslant \beta \|u,\dot{u}\|^2_{Y\times\mathcal{H}} \leqslant \beta \|(u,\dot{u},w)\|^2_X$$

for a constant β.

It remains to show that $(\lambda - G)$ has dense range for λ sufficiently large. If $(v,\dot{v},h)$ is orthogonal to the range then using the original inner product,

$$\langle \lambda u - \dot{u},v\rangle_Y + \langle \lambda\dot{u} - Au - Bw,\dot{v}\rangle_{\mathcal{H}} + \langle \lambda w - Cw,h\rangle_{\tilde{\mathcal{H}}} = 0,$$

for all $u \in D(\dot{A})$, $\dot{u} \in Y$, $w \in D(B)$. Taking $\dot{u} = \lambda u$ and $w = 0$ we get, since $(\lambda - A)$ is surjective, $\dot{v} = 0$. Then choosing $u,\dot{u} = 0$, and using the fact that $D(B)$ is a core for C, we find that $h = 0$ and finally $\dot{u} = 0$, $w = 0$ gives $v = 0$. □

The exact determination of $D(\overline{G})$ in examples is given by Dafermos [61] and in the appendix following. (It seems a little easier to show that $(\lambda - G)$ has dense range and to determine $D(\overline{G})$ later than to show directly that $(\lambda - \overline{G})$ is onto.)

APPENDIX (by C. Navarro): Existence and Uniqueness in the Cauchy problem for a linear thermoelastic material with memory.

This appendix presents a detailed example in which semigroup theory is applied to the problem of existence and uniqueness for evolutionary equations arising in linear thermoelasticity with memory effects.

More particularly, the evolutionary equations studied describe dynamical behaviour of a body whose constitutive assumptions are those of an inhomogeneous and anisotropic linear thermoelastic material with history of the fading memory type (cf. Coleman and Mizel [55]). This history depends upon the independent kinematic variables of displacement and temperature difference. The theory represents a linear approximation to that for a simple solid, apart from the fact that the functional dependence assumed for the heat flux vector is the same as in classical linear thermo-elasticity.

Asymptotic behaviour is not treated. However, such an analysis is unlikely to present any difficulty within the context of contraction semi-groups in Hilbert space and using techniques from topological dynamics (cf., Dafermos and Slemrod [66]†)

We proceed to formulate the problem. It is supposed that the body occupies a bounded region $\Omega \subset \mathbb{R}^n$ with smooth boundary $\partial\Omega$, and that the reference configuration is a natural state in which stress is zero and base temperature θ_0 is a strictly positive constant. Let $x \in \Omega$ be the position of a material point at time t, $u(x,t)$ the displacement and $\theta(x,t)$ the temperature difference from θ_0.

We assume that the Cauchy stress $\underset{\sim}{t}$ and specific entropy difference η are given by functionals depending upon both displacement and termperature difference history in the following form:

$$\underset{\sim}{t}(x,t) = g(x,0)\cdot\nabla u(x,t) - \theta(x,t)l(x,0)$$

$$+ \int_0^\infty \{g'(x,s)\cdot\nabla u(x,t-s) - l'(x,s)\theta(x,t-s)\}ds,$$

$$\rho(x)\eta(x,t) = l(x,0)\cdot\nabla u(x,t) + \rho(x)c(x,0)\theta(x,t)/\theta_0$$

$$+ \int_0^\infty \{l'(x,s)\cdot\nabla u(x,t-s) + \rho(x)c'(x,s)\theta(x,t-s)/\theta_0\}ds,$$

---

†A complete asymptotic stability analysis using a different method has been given elsewhere [197].

where $\rho(x)$ is the mass density in the natural state and $\partial a/\partial s$ is denoted by a'. The material functions $g(x,s)$, $l(x,s)$ and $c(x,s)$, $s \geqslant 0$ are the relaxation tensors of fourth, second and zero order respectively. We call the values $g(x,0)$, $l(x,0)$ and $c(x,0)$ of these quantities at $s=0$, the instantaneous elastic modulus, instantaneous stress-temperature tensor and instantaneous specific heat, respectively. In terms of Euclidean coordinates the system above states that

$$t_{ij}(x,t) = g_{ijkm}(x,0)u_{k,m}(x,t) - \theta(x,t)l_{ij}(x,0)$$

$$+ \int_0^\infty g'_{ijkm}(x,s)u_{k,m}(x,t-s)ds - \int_0^\infty l'_{ij}(x,s)\theta(x,t-s)ds,$$

and

$$\rho(x)\eta(x,t) = l_{ij}(x,0)u_{i,j}(x,t) + \rho(x)c(x,0)\theta(x,t)/\theta_0$$

$$+ \int_0^\infty l'_{ij}(x,s)u_{i,j}(x,t-s)ds + (\rho(x)/\theta_0)\int_0^\infty c'(x,s)\theta(x,t-s)ds$$

where the summation convention is employed.

The set of constitutive equations is completed by restricting attention to Fourier's law for the heat flux vector $q(x,t)$:

$$q(x,t) = -\kappa(x)\cdot\nabla\theta(x,t),$$

i.e.,

$$q_i(x,t) = -\kappa_{ij}(x)\theta_{,j}(x,t),$$

where $\kappa(x)$ is the thermal conductivity in the reference configuration.

Let v denote velocity and a superposed dot the partial time derivative. The local equation for balance of momentum is then

$$\operatorname{div} \underset{\sim}{t}(x,t) = \rho(x)\dot{v}(x,t)$$

while the linearised energy balance equation becomes

$$\theta_0\rho(x)\dot{\eta}(x,t) + \operatorname{div} q\,(x,t) = 0.$$

Substitution of $\underset{\sim}{t}$, η and q into these equations then yields the system of coupled equations for the linear theory of thermoelastic materials with memory:

$$\dot{v}(x,t) = \frac{1}{\rho(x)}\operatorname{div}\,(g(x,0)\cdot\nabla u(x,t) - \theta(x,t)l(x,0)$$

$$+ \int_0^\infty \{g'(x,s)\cdot\nabla u(x,t-s) - l'(x,s)\theta(x,t-s)\}ds)$$

$$\dot{\theta}(x,t) = \theta_0(\operatorname{div}\{\kappa(x)\cdot\nabla\theta(x,t)\}/\theta_0 - l(x,0)\cdot\nabla v(x,t)$$

$$+ \int_0^\infty \{l'(x,s)\cdot\nabla\dot{u}(x,t-s) + (\rho(x)/\theta_0)c'(x,s)\dot{\theta}(x,t-s)\}ds)/(\rho(x)c(x,0))$$

Obviously, we have supposed that the specific body force and the specific heat supply vanish identically, except possibly for $t < 0$.

The boundary conditions are assumed to be

$$u(x,t) = 0, \qquad \theta(x,t) = 0 \qquad \text{on} \quad \partial\Omega\times[0,\infty)$$

while the prescribed initial histories for the displacement and temperature difference are given by

$$u(x,-s) = w^0(x,-s), \qquad \theta(x,-s) = \alpha^0(x,-s), \qquad 0 \leqslant s < \infty, \quad x \in \bar{\Omega}.$$

We shall now state the main hypothesis on the material properties. First of all, we assume that the functions $\kappa(x)$, and $g(x,s)$, $l(x,s)$, $c(x,s)$ for fixed $s \geqslant 0$, are Lebesque measurable and essentially bounded on Ω. The associated norms are

$$\left\|\frac{\partial^i}{\partial s^i} g(s)\right\| = \operatorname*{ess.\,sup}_{x \in \Omega} \left|\frac{\partial^i}{\partial s^i} g(x,s)\right|, \qquad i = 1,2$$

$$\left\|\frac{\partial^i}{\partial s^i} c(s)\right\| = \operatorname*{ess.\,sup}_{x \in \Omega} \left|\frac{\partial^i}{\partial s^i} c(x,s)\right|,$$

where

$$\left|\frac{\partial^i}{\partial s^i} g(x,s)\right| = \sup_{|m|=1} \left|\frac{\partial^i}{\partial s^i} g_{pqrs}(x,s) m_{rs}\right|, \qquad |m| = (m_{pq} m_{pq})^{\frac{1}{2}},$$

are assumed to be continuous on $[0,\infty)$ and elements of $L_1(0,\infty)$. Moreover, we assume that

$$g_{ijkm}(x,s) = g_{kmij}(x,s) \qquad \text{a.e. on } \Omega$$

for all $s \geqslant 0$.

In addition we also postulate the following conditions:

(i) $0 < \rho_0 \leqslant \operatorname*{ess.\,inf}_{x \in \Omega} \rho(x)$,

(ii) $0 < c_0 \leqslant \operatorname*{ess.\,inf}_{x \in \Omega} c(x,0)$,

(iii) $\int_\Omega \nabla y(x) \cdot g(x,\infty) \cdot \nabla y(x) dV \geqslant g \int_\Omega y_{i,j}(x) y_{i,j}(x) dV$, for all $y \in C_0^\infty(\Omega)$

where g is a positive constant and $g(x,\infty) = \lim_{s \to \infty} g(x,s)$ is the equilibrium elastic modulus,

(iv) $\int_\Omega \nabla \beta(x) \cdot \kappa(x) \cdot \nabla \beta(x) dV \geqslant k \int_\Omega \beta_{,i}(x) \beta_{,i}(x) dV$, for all $\beta \in C_0^\infty(\Omega)$

where k is a positive constant,

(v) $\int_\Omega \nabla y(x) \cdot g''(x,s) \cdot \nabla y(x) dV \geqslant g_2(s) \int_\Omega y_{i,j}(x) y_{i,j}(x) dV$,

for all $y \in C_0^\infty(\Omega)$

where $0 \leqslant s < \infty$ and $g_2(s) \geqslant 0$,

(vi) $$-\int_\Omega c''(x,s)\beta^2(x)dV \geqslant c_2(s)\int_\Omega |\beta(x)|^2 dV, \quad \text{for all} \quad \beta \in C_0^\infty(\Omega)$$

where $0 \leqslant s < \infty$ and $c_2(s) \geqslant 0$,

(vii) For all $s \geqslant 0$,

$$\| l'(s) \| = \operatorname*{ess.\ sup}_{x \in \Omega} [\, l'_{pq}(x,s) l'_{pq}(x,s)\,]^{\frac{1}{2}} \in L_1(0,\infty)$$

$$\| l''(s) \| \leqslant \left(\frac{\rho_0}{\theta_0}\right)^{\frac{1}{2}} [\, c_2(s)\,]^{\frac{1}{2}} [\, g_2(s)\,]^{\frac{1}{2}}.$$

Remarks 1. Since $\lim_{s \to \infty} \| l'(s) \| = 0$, condition (vii) and the Cauchy-Schwarz inequality imply

$$\| l'(s) \| \leqslant (\rho_0/\theta_0)^{\frac{1}{2}} [\, c_1(s)\,]^{\frac{1}{2}} [\, g_1(s)\,]^{\frac{1}{2}}, \qquad 0 \leqslant s < \infty, \tag{1}$$

where

$$c_1(s) = \int_s^\infty c_2(s')ds' \quad \text{and} \quad g_1(s) = \int_s^\infty g_2(s')ds'.$$

Then,

$$\int_\Omega c'(x,s)\beta^2(x)dV \geqslant c_1(s)\int_\Omega |\beta(x)|^2 dV, \quad \text{for all} \quad \beta \in C_0^\infty(\Omega) \tag{2}$$

$$-\int_\Omega \nabla y(x)\cdot g'(x,s)\cdot\nabla y(x)dV \geqslant g_1(s)\int_\Omega y_{i,j}(x)y_{i,j}(x)dV, \quad \text{for all} \quad y \in C_0^\infty(\Omega). \tag{3}$$

2. Condition (iv) is clearly related to the heat conduction inequality. Assumptions (v), (vi) and (vii) are motivated by sufficient conditions for the internal dissipation inequality to be satisfied

(cf. Navarro [197]).

Let now $w(x,t,s)$ and $\alpha(x,t,s)$ denote the displacement and temperature difference histories, i.e., $w(x,t,s) = u(x,t-s)$ and $\alpha(x,t,s) = \theta(x,t-s)$, $0 \leqslant s \leqslant \infty$.

<u>Definition</u> Consider the elements $(u(x),v(x),\theta(x),w(x,s),\alpha(x,s))$. We shall denote by X the Hilbert space obtained as the completion of the set

$$(u,v,\theta,w,\alpha) \in C_0^\infty(\Omega) \times C_0^\infty(\Omega) \times C_0^\infty(\Omega) \times C^\infty([0,\infty);H_0^1(\Omega)) \times C^\infty([0,\infty);H_0^1(\Omega))$$

under the norm induced by the inner product

$$\begin{aligned}\langle (u,v,\theta,w,\alpha),(\bar{u},\bar{v},\bar{\theta},\bar{w},\bar{\alpha}) \rangle = &\int_\Omega \{\nabla u \cdot g(\infty) \cdot \nabla \bar{u} + \rho v\bar{v} + \frac{\rho c(0)}{\theta_0} \theta\bar{\theta}\} dV \\ &- \int_\Omega \int_0^\infty \{[\nabla u - \nabla w(s)] \cdot g'(s) \cdot [\nabla \bar{u} - \nabla \bar{w}(s)] \\ &+ \bar{\alpha}(s) l'(s) \cdot [\nabla u - \nabla w(s)] \\ &+ \alpha(s) l'(s) \cdot [\nabla \bar{u} - \nabla \bar{w}(s)] - \frac{\rho}{\theta_0} c'(s) \alpha(s) \bar{\alpha}(s)\} ds dV\end{aligned} \tag{4}$$

where dependence on x is omitted for convenience.

<u>Remark</u> It is obvious from (i), (ii) and (iii) that the first of the two integrals appearing in the expression for $\langle (u,v,\theta,w,\alpha),(u,v,\theta,w,\alpha) \rangle$ is non-negative, whereas from (1) - (3) the second is bounded below by

$$\int_0^\infty ([g_1(s)]^{\frac{1}{2}} \int_\Omega |\nabla u - \nabla w(s)|^2 dV - [(\rho_0/\theta_0)c_1(s)]^{\frac{1}{2}} \int_\Omega |\theta|^2 dV)^2 ds \geqslant 0.$$

Thus, we see that (4) is indeed an inner product.

We now transform our boundary-initial history value problem to an initial value problem in the Hilbert space X. To do this we define the operator

$$G\begin{pmatrix} u \\ v \\ \theta \\ w \\ \alpha \end{pmatrix} = \begin{pmatrix} v \\ \frac{1}{\rho}\operatorname{div}\{g(0)\cdot\nabla u - \theta 1(0) + \int_0^\infty [\, g'(s)\cdot\nabla w(s) - 1'(s)\alpha(s)\,]ds\} \\ (\theta_0/\rho c(0))\{-1'(0)\cdot\nabla u - 1(0)\cdot\nabla v + \operatorname{div}(\kappa\cdot\nabla\theta)/\theta_0 - (\rho c'(0)/\theta_0)\theta - \\ - \int_0^\infty [\, 1''(s)\cdot\nabla w(s) + (\rho/\theta_0)c''(s)\alpha(s)\,]ds\} \\ -w'(s) \\ -\alpha'(s) \end{pmatrix}$$

with domain D(G) given by

$$D(G) = \{(u,v,\theta,w,\alpha) \in X \ \ G\begin{pmatrix} u \\ v \\ \theta \\ w \\ \alpha \end{pmatrix} \in X \ \text{ and } \ w(0) = u, \ \ \alpha(0) = \theta \ \text{ for } \ x \in \Omega\}.$$

Thus, we obtain the abstract evolutionary equation

$$\frac{d}{dt}\begin{pmatrix} u(t) \\ v(t) \\ \theta(t) \\ w(t) \\ \alpha(t) \end{pmatrix} = G\begin{pmatrix} u(t) \\ v(t) \\ \theta(t) \\ w(t) \\ \alpha(t) \end{pmatrix}, \quad \begin{pmatrix} u(0) \\ v(0) \\ \theta(0) \\ w(0) \\ \alpha(0) \end{pmatrix} = \begin{pmatrix} w^0(0) \\ v^0 \\ \alpha^0(0) \\ w^0 \\ \alpha^0 \end{pmatrix}.$$

We now state the existence and uniqueness result for the equation above.

<u>Theorem</u> The operator G is the infinitesimal generator of a C_0 contractive semigroup on X.

<u>Proof</u>. We first note that D(G) is dense in X. Furthermore, and after a

lengthy but straightforward calculation, we obtain for $(u,v,\theta,w,\alpha) \in D(G)$, the following inequality analogous to that in the theorem preceding this appendix;

$$\begin{aligned}
\langle G(u,v,\theta,w,\alpha),(u,v,\theta,w,\alpha)\rangle = & -\tfrac{1}{2}\int_\Omega\int_0^\infty \{[\nabla u - \nabla w(s)]\cdot g''(s)\cdot[\nabla u - \nabla w(s)] \\
& - (\rho/\theta_0)c''(s)[\theta - d(s)]^2 \\
& - 2[\theta - \alpha(s)]\, l''(s)\cdot[\nabla u - \nabla w(s)]\}dsdV \\
& - (1/\theta_0)\int_\Omega \nabla\theta\cdot\kappa\cdot\nabla\theta dV \\
\leqslant & - k\int_\Omega |\nabla\theta|^2 dV - \int_0^\infty\{[g_2(s)]^{\frac{1}{2}}\int_\Omega |\nabla u - \nabla w(s)|^2 dV \\
& - [(\rho/\theta_0)c_2(s)]^{\frac{1}{2}}\int_\Omega |\theta - \alpha(s)|^2 dV\}^2 ds \\
\leqslant & \; 0,
\end{aligned}$$

where we have made use of integration by parts and conditions (iv), (v), (vi) and (vii). Therefore, we conclude that G is a dissipative operator.

Next we show that the range condition $R(I - G) = X$ is satisfied. Then, the theorem will be proved as a consequence of the second Corollary to the Hille-Yosida theorem (see Section 5).

Assume that $(\overline{u},\overline{v},\overline{\theta},\overline{w},\overline{\alpha}) \in X$. Then we must show that the system of equations

$$(I - G)\begin{pmatrix} u \\ v \\ \theta \\ w \\ \alpha \end{pmatrix} = \begin{pmatrix} \overline{u} \\ \overline{v} \\ \overline{\theta} \\ \overline{w} \\ \overline{\alpha} \end{pmatrix} \tag{5}$$

has a solution (u,v,θ,w,α) in $D(G)$. The last two equations of (5) are

$$w(s) + w'(s) = \overline{w}(s)$$
$$\alpha(s) + \alpha'(s) = \overline{\alpha}(s) \tag{6}$$

and can be invented to give

$$w(s) = e^{-s}\left(u + \int_0^s e^{\xi}\overline{w}(\xi)d\xi\right), \qquad \alpha(s) = e^{-s}\left(\theta + \int_0^s e^{\xi}\overline{\alpha}(\xi)d\xi\right). \tag{7}$$

Introducing (7) into the second and third equations of (5), and using the first equation, we get the system

$$A'\begin{pmatrix} u \\ \theta \end{pmatrix} = \begin{pmatrix} A & F \\ C & D \end{pmatrix}\begin{pmatrix} u \\ \theta \end{pmatrix} = \begin{pmatrix} h_1 \\ h_2 \end{pmatrix}, \tag{8}$$

where

$$Au = u - (1/\rho)\operatorname{div}\{[\, g(0) + \int_0^\infty g'(s)e^{-s}ds \,]\cdot\nabla u\},$$

$$F\theta = (1/\rho)\operatorname{div}\{[\, l(0) + \int_0^\infty l'(s)e^{-s}ds \,]\theta\},$$

$$Cu = (\theta_0/\rho c(0))(l(0) + \int_0^\infty l'(s)e^{-s}ds)\cdot\nabla u,$$

$$D\theta = \theta - (\theta_0/\rho c(0))\{\frac{1}{\theta_0}\operatorname{div}(\kappa\nabla\theta) - \frac{\rho}{\theta_0}(\int_0^\infty c'(s)e^{-s}ds)\theta\},$$

$$h_1 = \overline{v} + \overline{u} + (1/\rho)\operatorname{div}\{\int_0^\infty g'(s)e^{-s}\cdot\int_0^s e^{\xi}\nabla\overline{w}(\xi)d\xi ds - \int_0^\infty l'(s)e^{-s}\int_0^s e^{\xi}\overline{\alpha}(\xi)d\xi ds\},$$

$$h_2 = \overline{\theta} + (\theta_0/\rho c(0))\{l(0)\cdot\nabla\overline{u} + \int_0^\infty [\, l'(s)\cdot\nabla\overline{w}(s) + (\rho/\theta_0)c'(s)\overline{\alpha}(s)\,]ds$$
$$- \int_0^\infty l'(s)e^{-s}\cdot\int_0^s e^{\xi}\nabla\overline{w}(\xi)d\xi ds - (\rho/\theta_0)\int_0^\infty c'(s)e^{-s}\int_0^s e^{\xi}\overline{\alpha}(\xi)d\xi ds\}.$$

Concerning (8) we have the following:

<u>Lemma</u> There exists a unique solution $(u,\theta) \in H_0^1 \times H_0^1$ to the system (8).

<u>Proof.</u> If $\langle \cdot,\cdot \rangle$ are L_2 inner products conveniently weighted, the bilinear form $B[(u,\theta),(\hat{u},\hat{\theta})] = \langle A'(u,\theta),(\hat{u},\hat{\theta})\rangle$ clearly determines a norm equivalent to $\|(u,\theta)\| = \|u\|_{H_0^1} + \|\theta\|_{H_0^1}$, since

$$B[(u,\theta),(u,\theta)] = \langle u, Au + F\theta\rangle + \langle \theta, Cu + D\theta\rangle$$

$$= \int_\Omega \{\rho u^2 + \nabla u\cdot(g(0) + \int_0^\infty g'(s)e^{-s}ds)\cdot\nabla u + (1/\theta_0)\nabla\theta\cdot\kappa\cdot\nabla\theta$$

$$+ (\rho/\theta_0)(\int_0^\infty c'(s)e^{-s}ds)\theta^2\}dV.$$

All we need to prove now is that $(h_1,h_2) \in H^{-1}\times H^{-1}$ with norm $\|(h_1,h_2)\| = \|h_1\|_{H^{-1}} + \|h_2\|_{H^{-1}}$, where

$$\|h_i\|_{H^{-1}} = \sup_{y\in H_0^1} \frac{\left|\int_\Omega y h_i dV\right|}{\|y\|_{H_0^1}}, \qquad i = 1,2.$$

First, since

$$\frac{\partial}{\partial s}\int_s^\infty g'(\xi)e^{-\xi}d\xi = -g'(s)e^{-s},$$

integration by parts yields the identity

$$\int_0^\infty g'(s)e^{-s}\cdot\int_0^s e^{\xi}\nabla\overline{w}(\xi)d\xi ds = \int_0^\infty [\int_s^\infty g'(\xi)e^{-\xi}d\xi\,]e^{s}\cdot\nabla\dot{w}(s)ds,$$

and, in the same way,

$$\int_0^\infty l'(s)e^{-s}\int_0^s l^{\xi}\overline{\alpha}(\xi)d\xi ds = \int_0^\infty [\int_s^\infty l'(\xi)e^{-\xi}d\xi\,]e^{s}\overline{\alpha}(s)ds.$$

Next, choose any $y \in H_0^1(\Omega)$. Then,

$$\left|\int_\Omega y h_1 dV\right| = \left|\int_\Omega \{(\bar{v}+\bar{u})y - \nabla y\cdot\int_0^\infty(\int_s^\infty g'(\xi)e^{-\xi}d\xi)\cdot e^s\nabla\bar{w}(s)ds\right.$$

$$\left. + \nabla y\cdot\int_0^\infty(\int_s^\infty 1'(\xi)e^{-\xi}d\xi)e^s\bar{\alpha}(s)ds\}dV\right|$$

$$\leqslant \int_0^\infty\{(\int_\Omega \nabla y\cdot[-\int_s^\infty g'(\xi)e^{-\xi}d\xi]e^s\cdot\nabla y dV)^{\frac{1}{2}} \times$$

$$\times (\int_\Omega \nabla\bar{w}(s)\cdot[-\int_s^\infty g'(\xi)e^{-\xi}d\xi]e^s\cdot\nabla\bar{w}(s)dV)^{\frac{1}{2}} + \|\int_s^\infty 1'(\xi)e^{-\xi}d\xi\|e^s(\int_\Omega|\nabla y|^2dV)^{\frac{1}{2}} \times$$

$$\times (\int_\Omega|\bar{\alpha}(s)|^2dV)^{\frac{1}{2}}\}ds + \|\bar{v}+\bar{u}\|_{L_2}\|y\|_{H_0^1}$$

$$\leqslant \|y\|_{H_0^1}\{\|g(0)-g(\infty)\|^{\frac{1}{2}}([-\int_0^\infty\int_\Omega \nabla\bar{w}(s)\cdot g'(s)\cdot\nabla\bar{w}(s)dVds]^{\frac{1}{2}}$$

$$+ [(\rho_0/\theta_0)\int_0^\infty\int_\Omega c'(s)\bar{\alpha}^2(s)dVds]^{\frac{1}{2}}) + \|\bar{v}+\bar{u}\|_{L_2}\}.$$

Here we have used the Cauchy-Schwarz and Young inequalities, the estimate

$-\int_0^\infty g'(\xi)e^{-\xi}d\xi \leqslant -g'(s)e^{-s}$ and the inequality

$$\|\int_0^\infty 1'(\xi)e^{-\xi}d\xi\|e^s \leqslant e^s(\rho_0/\theta_0)^{\frac{1}{2}}\int_s^\infty[c_1(\xi)g_1(\xi)]^{\frac{1}{2}}e^{-\xi}d\xi$$

$$\leqslant e^s(\int_s^\infty \rho_0\theta_0^{-1}c_1(\xi)e^{-\xi}d\xi)^{\frac{1}{2}}(\int_s^\infty g_1(\xi)e^{-\xi}d\xi)^{\frac{1}{2}}$$

$$\leqslant e^s(\rho_0\theta_0^{-1}c_1(s)e^{-s})^{\frac{1}{2}}(g_1(s)e^{-s})^{\frac{1}{2}}$$

$$= (\rho_0\theta_0^{-1}c_1(s)g_1(s))^{\frac{1}{2}},$$

which is implied by (1) and the fact that $c_1(s)$, $g_1(s)$ are positive, monotone decreasing functions. Finally, we have made use of (2) and (3). Therefore, we conclude that $h_1 \in H^{-1}(\Omega)$.

It can be shown in a similar manner that $h_2 \in H^{-1}(\Omega)$. Hence, Riesz's theorem implies the existence of a unique pair $(u,\theta) \in H_0^1(\Omega) \times H_0^1(\Omega)$.

Going back to the proof of the theorem, and considering the first equation of (5), $u - v = \overline{u}$, we see that $v \in H_0^1(\Omega)$ by the Lemma. On the other hand, the equations

$$\int_0^\infty\int_\Omega \{\nabla w(s)\cdot g'(s)\cdot\nabla w(s) + \nabla w(s)\cdot g'(s)\cdot\nabla w'(s)\}dVds$$

$$= \int_0^\infty\int_\Omega \nabla w(s)\cdot g'(s)\cdot\nabla\overline{w}(s)dVds,$$

$$\int_0^\infty\int_\Omega \rho\theta_0^{-1}\{c'(s)\alpha^2(s) + c'(s)\alpha(s)\alpha'(s)\}dVds = \int_0^\infty\int_\Omega \rho\theta_0^{-1}c'(s)\alpha(s)\overline{\alpha}(s)dVds$$

follow immediately from (6). The use in the expressions above of the Cauchy-Schwarz and Young inequalities, together with integration by parts and assumptions (v), (vi), yield the estimates

$$-\int_0^\infty\int_\Omega \nabla w(s)\cdot g'(s)\cdot\nabla w(s)dVds \leqslant -\int_0^\infty\int_\Omega \nabla\overline{w}(s)\cdot g'(s)\cdot\nabla\overline{w}(s)dVds$$

$$-\int_\Omega \nabla u\cdot g'(0)\cdot\nabla u dV$$

and

$$\int_0^\infty\int_\Omega \rho\theta_0^{-1}c'(s)\alpha^2(s)dVds \leqslant \int_0^\infty\int_\Omega \rho\theta_0^{-1}c'(s)\overline{\alpha}^2(s)dVds + \int_\Omega \rho\theta_0^{-1}c'(0)\theta^2 dV.$$

Furthermore, condition (1) and standard inequalities imply

$$2\left|\int_0^\infty\int_\Omega \alpha(s)1'(s)\cdot\nabla w(s)dVds\right| \leqslant \int_0^\infty\int_\Omega \{\rho\theta_0^{-1}c'(s)\alpha^2(s)$$

$$-\nabla w(s)\cdot g'(s)\cdot\nabla w(s)\}dVds$$

and this estimate, with the two previous ones and (6), shows that

$$-\int_0^\infty\int_\Omega \{\nabla w'(s)\cdot g'(s)\cdot\nabla w'(s) + 2\alpha'(s)l'(s)\cdot\nabla w'(s)$$

$$- \rho\theta_0^{-1}c'(s)\alpha'(s)^2\} dV ds < \infty.$$

We conclude that $(u,v,\theta,w,\alpha) \in D(G)$, i.e., $R(I-G) = X$, and the theorem is proved. □

Remark Introduction of a specific body force and specific heat supply to study the forced problem causes us no additional difficulty as regards existence and uniqueness (see Section 5, variation of constants formula).

8 LINEAR AND LOCAL NONLINEAR ELASTOSTATICS

This section discusses for completeness the existence and uniqueness theory for linear elastostatics and for nonlinear elastostatics with data near that of a given solution. The emphasis is just slightly different from that given by, for example, Fichera [83].

The linearized equations under consideration may be written (see Section 4)

$$\operatorname{div}(a\cdot\nabla u)(x) = f(x), \qquad x \in \Omega$$

where $f = -\rho b - \operatorname{div} \underset{\sim}{t}$, b is the body force and $\underset{\sim}{t}$ is the Cauchy stress in the configuration we are linearizing about and $a = \underset{\sim}{t}\otimes\delta + c$ is the corresponding elasticity tensor (assumed sufficiently smooth). We assume the boundary conditions are either displacement or traction. They may be assumed to be homogeneous. (If they are not, match them up arbitrarily to a displacement $\tilde{u}$ and replace f by $f + \operatorname{div}(a\cdot\nabla\tilde{u})$ and u by $u - \tilde{u}$.)

We let $A(u) = \operatorname{div}(a\cdot\nabla u)$ be the linear operator in $L_2(\Omega)$ with domain

$H^2(\Omega)$ with the relevant boundary conditions imposed.

We make the assumptions:

1. a is hyperelastic, i.e., $a_{ijkl} = a_{jilk}$ and so A is a symmetric operator.

2. a is strongly elliptic, i.e., there is an $\varepsilon > 0$ such that

$$a_{ijkl}(x)\xi_i\xi_k\eta_j\eta_l \geqslant \varepsilon|\xi|^2|\eta|^2,$$

for all $x \in \Omega$ and $\xi,\eta \in \mathbb{R}^n$. (See the previous section.)

As mentioned in the previous section, these two assumptions imply that the elliptic estimates hold, i.e., if $u \in D(A)$ then for each $s \geqslant 2$ and $1 < p < \infty$ there is a constant C such that

$$\|u\|_{W^{s,p}} \leqslant C(\|Au\|_{W^{s-2,p}} + \|u\|_{L^2}),$$

(if $Au \in W^{s-2,p}$). ....see Morrey [191] or Agmon-Douglis and Nirenberg [3].[†]
Strong ellipticity also implies (from the elliptic estimates and Rellich's theorem) that

(i) ker A is finite dimensional

(ii) $(\lambda - A)^{-1}$ is compact for λ sufficiently large,

(iii) the spectrum of A is discrete and each eigenvalue has finite multiplicity.

Of more concern to us is:

[†]One has to verify that the boundary conditions satisfy the complementing (or Lopatinsky-Shapiro) conditions. This is clear for the displacement problem and may also be verified for the traction problem.

__Theorem__ $L_2(\Omega) = \text{Range }(A) \oplus \text{Ker }(A)$. ($L_2$ orthogonal direct sum.)

__Proof.__ First of all, for any selfadjoint operator A in a Hilbert space $\mathcal{H}$ we have the formally obvious fact

$$\mathcal{H} = \overline{\text{Range }(A)} \oplus \overline{\text{Ker }(A)}.$$

(See, for instance, Yosida [261]). In our case, ellipticity shows that $\text{Ker}\, A$ is closed because it is finite dimensional and, more significantly, the elliptic estimates show that the range is closed as well. This yields the stated theorem. □

If we write the decomposition as

$$f = Au + g, \qquad g \in \text{Ker}\, A$$

then $A^2 u = f$ so Au is as smooth as f. From this and the elliptic estimates we see that we get __regularity__ of the decomposition, i.e., the more delicate decomposition

$$W^{s,p}(\Omega) = [\text{Range}\,(A) \cap W^{s,p}(\Omega)] \oplus [\text{Ker}\,(A) \cap W^{s,p}(\Omega)],$$

where again the sum is L_2 orthogonal.

The above are __general facts__ for elliptic boundary value problems usually referred to as the __Fredholm alternative__. (See Wells [254] and for the case of the Hodge theorem, Morrey [191]). In the non-self adjoint case, the result reads†

$$W^{s,p} = [\text{Range}\,(A) \cap W^{s,p}] \oplus [\text{Ker}\,(A^*) \cap W^{s,p}].$$

---

†These results have wide applicability in geometry in addition to partial differential equations. See, for example Fischer and Marsden [86] and the references therein.

From these we can deduce immediately the main result for our boundary value problem.

Theorem Let $f \in L_2(\Omega)$. Then there exists a $u \in D(A)$ such that

$$Au = f$$

if and only if

$$\langle f,h \rangle = 0 \quad \text{for all} \quad h \in \ker(A)$$

In this case (i) u is unique up to addition of elements in $\ker A$;

(ii) if f is of class $W^{s,p}$, so is u (up to and including the boundary).

In particular, note that if A is stable then $\ker A = \{0\}$, so one has existence and uniqueness. As discussed in the previous section this can be verified, by the use of Korn's inequalities, for classical elasticity if c_{ijkl} is positive definite. (If $\underset{\sim}{t}$ is sufficiently small, similar results may be deduced.) The calculation of $\mathrm{Ker}\, A$ however can be done directly in this case and Korn's inequalities are not needed. (If $u_{i,j} + u_{j,i} = 0$, u must be an infinitesimal Euclidean motion; if $u = 0$ on $\partial\Omega$, u must vanish; if $c\cdot\nabla u = 0$ on $\partial\Omega$, u is unrestricted.)

Corollary For the equations of classical elasticity, assume that c_{ijkl} is positive definite: i.e., there is an $\varepsilon > 0$ such that

$$c_{ijkl}e_{ij}e_{kl} \geqslant \varepsilon \| e \|^2$$

for all symmetric e_{ij}. Then

(i) (Displacement problem) For any $f \in L_2(\Omega)$

$$\mathrm{div}(c\cdot\nabla u) = f$$

has a unique solution $u \in H^2(\Omega)$, $u = 0$ on $\partial\Omega$. If $f \in W^{s,p}$, then $u \in W^{s+2,p}$ for $s \geq 0$, $1 < p < \infty$.

(ii) (Traction problem) The equation

$$\operatorname{div}(\mathbb{c} \cdot \nabla u) = f, \qquad f \in L_2(\Omega),$$

with $\mathbb{c} \cdot \nabla u = 0$ on $\partial\Omega$ has a solution $u \in H^2(\Omega)$ if and only if

$$\int_\Omega f(x) \cdot (a + bx)\,dx = 0$$

where a is any constant vector and b is any skew symmetric matrix. If this holds, u is unique up to the addition of a term of the form $a + bx$, b a skew matrix. If $f \in W^{s,p}$, $s \geq 0$, $1 < p < \infty$, then $u \in W^{s+2,p}$.

We now discuss the nonlinear problem following the methods outlined in Section 4. First we consider the displacement problem. Suppose we are in a situation in which the linearized problem about a given configuration $\overset{\circ}{\phi}$ has a unique solution (e.g., stability holds). Then, by the composition theorem in Sobolev spaces (Section 4), the map

$$\phi \longmapsto \mathrm{DIV}(\hat{T}(F)) + \rho_0 B$$

will be smooth provided ϕ is of class $W^{s,p}$, $s \geq 2$ and $s > \frac{n}{p} + 1$, with the range space $W^{s-2,p}$. Moreover, its linearization at a solution $\overset{\circ}{\phi}$ will be an isomorphism. We can conclude from the inverse function theorem that

(i) the solution $\overset{\circ}{\phi}$ is isolated (locally unique) in $W^{s,p}$;

(ii) if the data (B or the boundary conditions) are close enough to those of $\overset{\circ}{\phi}$, the <u>perturbed problem has a locally unique solution</u>.

Of course, if parameters like the loads or boundary conditions are varied over a larger range, we expect the linearization to cease being an isomorphism and bifurcations to occur. Nevertheless, in analyzing the problem, use of the inverse function theorem is a basic first step. (The bifurcation results are an entirely different story not entered into here.)

The above results go back to Stoppelli [235]. We refer also to John [135] and for additional references to Wang and Truesdell [252].

However there are some words of caution we would like to give by way of an example (see Ball, Knops and Marsden [16]). In the existence theory for minimizers in the calculus of variations, the spaces $W^{1,p}$ play an important role. This is the case in elasticity as well, in view of the fundamental results of Ball [12] (see also Ball's article in volume 1 of these proceedings). Unfortunately the inverse function theorem does not work in $W^{1,p}$, even when the linearized problem about the equilibrium solution is stable. We consider the displacement problem in one dimension, writing u for the nonlinear displacement: $u = \phi$-identity. On $[0,1]$ we consider a stored energy function $W(u_x)$, no external forces and boundary conditions $u(0) = u(1) = 0$. Assume W is smooth and let $p_- < 0 < p_+$ be such that

$$W'(p_-) = W'(0) = W'(p_+)$$

and

$$W''(0) > 0.$$

(See figure 10.)

In $W^{2,p}$ (with the boundary conditions $u(0) = 0$, $u(1) = 0$), the trivial solution $u_0 \equiv 0$ is isolated because the map

$$u \longmapsto W(u_x)_x$$

from $W^{2,p}$ to L^p is smooth and its derivative at u_0 is the linear operator

$$v \longmapsto W''(p_0)v_{xx}$$

which is an isomorphism. Therefore, by the inverse function theorem, zeros of $W(u_x)_x$ are isolated, as above.

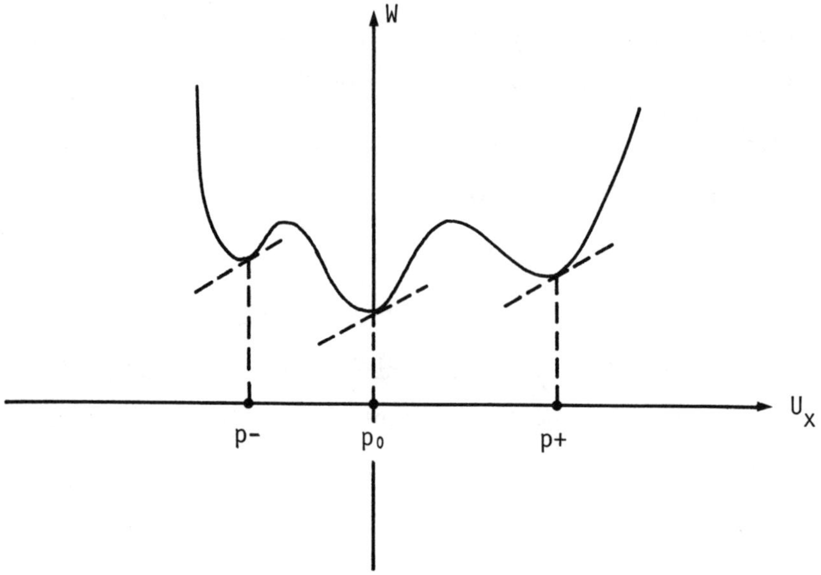

Fig. 10.

The second variation of $V(u) = \int_0^1 W(u_x)dx$ is positive definite (relative to the $H^1 = W^{1,2}$ topology) at u_0 because if v is in $W^{1,2}$ and vanishes at $x = 0,1$, then

$$\frac{d^2}{d\varepsilon^2}V(u_0 + \varepsilon v)_{|\,\varepsilon=0} = W''(p_0)\int_0^1 {v_x}^2 dx$$

$$\geqslant c\,\|v\|^2_{W^{1,2}}.$$

Now we show that u_0 is not isolated in $W^{1,p}$. Given $\varepsilon > 0$, let

$$u_\varepsilon(x) = \begin{cases} p_+ x & \text{for } 0 \leqslant x \leqslant \varepsilon, \\ p_+\varepsilon + p_-(x-\varepsilon) & \text{for } \varepsilon \leqslant x \leqslant \dfrac{p_- - p_+}{p_-}\varepsilon, \\ 0 & \text{for } \dfrac{p_- - p_+}{p_-}\varepsilon, \leqslant x \leqslant 1 \end{cases}$$

(See Fig. 11)

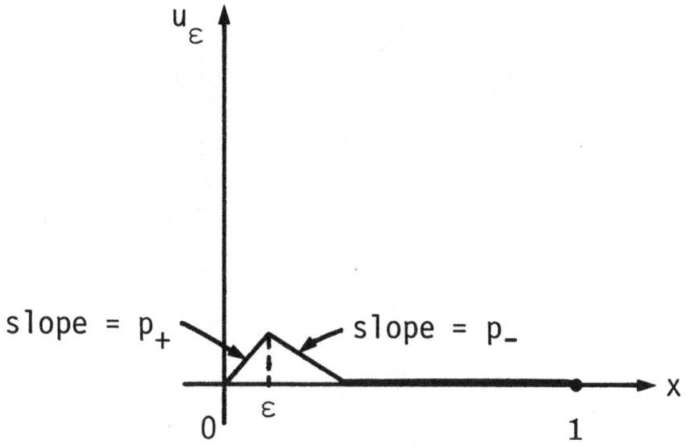

Fig. 11.

Since $W'(u_{\varepsilon x})$ is constant, each u_ε is an extremal. Also

$$\int_0^1 |u_{\varepsilon x} - u_{0x}|^p \, dx = \varepsilon |p_+|^p + \varepsilon |p_-|^p |p_+|^p$$

which tends to zero as $\varepsilon \longrightarrow 0$. Thus u_0 is not isolated in $W^{1,p}$.

Remarks 1. If $W(p_-) = W(p_+) = W(0)$ and if $W(p) \geqslant W(0)$ for all p, the same argument shows that there are absolute minima of J arbitrarily close to u_0 in $W^{1,p}$.

2. Phenomena like this seem to have first been noticed by Weierstrass. See Bolza [28], footnote 1, p.40.

Next we briefly discuss the traction problem. Here the situation, even locally, is much more interesting, because of the kernel which A always has, namely the infinitesimal Euclidean motions described above.

We consider an unstressed equilibrium state and then for given loads B_0 and τ_0, we ask if, for θ small enough the traction problem with loads $B = \theta B_0$ and $\tau = \theta\tau_0$ has a solution near equilibrium. These loads must satisfy the necessary condition

$$\int_{\partial\Omega}\tau dA + \int_{\Omega}\rho_0 B dV = 0 \tag{1}$$

as observed in Section 3. According to Da Silvas theorem (Truesdell and Noll [248], p.128) we can make a rotation of the reference configuration so that the second necessary condition for a solution always holds:

$$\int_{\partial\Omega}(X\times\tau(x))dA + \int_{\Omega}\rho_0 X\times B(X)dV = 0. \tag{2}$$

(Since the loads are not rotated, the reader is cautioned that this actually changes the problem.) A load (B,τ) is called equilibrated if both these equations (1) and (2) hold. An equilibrated load is said to have a vector e as an axis of equilibrium if any rotation of the body about this axis, keeping the loads fixed, maintains the equilibration conditions (1) and (2). For instance, axes of symmetry are axes of equilibrium, but the latter need not be axes of symmetry; (see Capriz and Guidugli [33] for an example). Let us write $l = (B,\tau)$ for the load.

The solution $x = \phi(X)$ must also be such that the moments are balanced:

$$\int_{\partial\Omega}x\times\tau(X)dA + \int_{\Omega}\rho_0 x\times B(X)dV = 0$$

a compatibility condition of Signorini that cannot be a priori assigned since ϕ is unknown.

Theorem (Stoppelli [235]) Suppose l is equilibrated, has no axis of equilibrium, and the linearized problem is strongly elliptic (see the preceding theorem). Then if $\phi = \mathrm{Id}$ is an unstressed state, and λ is sufficiently small[†] there is a unique solution of the traction problem near Id which fixes a given point, say $\phi(0) = 0$.

Sketch of Proof. (Van Buren [251]) The proof is a somewhat clever use of the inverse function theorem.

Under appropriate Sobolev or Hölder differentiability assumptions, let

$$\mathbb{B} = \{\text{loads } l \text{ which are equilibrated on } \Omega\},$$

$$\mathbb{D}_1 = \{\text{displacements } u \text{ near } 0 \text{ such that } u(0) = 0\},$$

$$\mathbb{D}_2 = \{u \in \mathbb{D}_1 \text{ such that } \nabla u(0) \text{ is symmetric}\}.$$

Let $\Phi(u) = (-\mathrm{DIV}\, T(u), T\cdot n|_{\partial\Omega})$, so we are trying to solve

$$\Phi(u) = \lambda l.$$

The inverse function theorem does not apply as things stand because of rotational ambiguities; the linearized problem has a three dimensional kernel, namely the skew-matrices, i.e., infinitesimal rotations. To get around this, one uses

Stoppelli's Lemma Suppose l has no axis of equilibrium. Then there is a neighbourhood N of Id in $\mathbb{D}_2$ and a unique map $\overline{Q} : N \longrightarrow SO(3)$ such that

$$\Phi(u) - \overline{Q}(u)l$$

is equilibrated; $\overline{Q}(\mathrm{Id}) = \mathrm{Id}$ and $D\overline{Q}(\mathrm{Id}) = 0$.

This is a direct application of the inverse function theorem; the three conditions of momentum equilibration uniquely determine $\overline{Q}$. To avoid

---

[†]The smallness of λ depends on l.

special kernels, we need no axis of equilibrium, as is intuitively evident.

Now define

$$\Psi_\lambda : \text{nbhd of Id in } \mathbb{D}_2 \longrightarrow \mathbb{B}_2,$$

$$u \longmapsto \Phi(u) - \overline{Q}(u)\lambda l .$$

Now $D\Psi_\lambda(\mathrm{Id})$ are the equations of the linearized traction problem; this map is a linear isomorphism from $\mathbb{D}_2$ to $\mathbb{B}_2$ (on these spaces the rotational indeterminacy is gone). So the inverse function theorem enables us to solve $\Psi_\lambda(u) = 0$ provided we can get near zero; this is done by taking λ small. This is slightly tricky because Ψ_λ itself depends on l, so the estimates involved have to be looked at explicitly.

By material frame indifference, the solution to the original problem is $\overline{Q}(u)^T u$. □

Notice that the rotational ambiguity present at the linearized level is "absent" at the nonlinear level and this is one reason for the delicacy of the proof.

The only references in English where results for the case of an axis of equilibrium are discussed are Grioli [105] and Capriz and Guidugli [33]. The only reference for the proofs (known to us) is Stoppelli [236]. We will discuss the problem from a completely different point of view in Section 11.

9 DYNAMICAL SYSTEMS AND HAMILTONIAN STRUCTURES

The abstract set up developed in this Section is designed for classical (smooth) solutions of the dynamical equations of elasticity. This means, in effect, that we are concerned with solutions, maximally developed in time, but before shocks develop. While shock structures are important, so is the dynamical system in the pre-shock stage. In addition, one should distinguish singularities in time which are "essential" in the sense that by enlarging the space and possibly extending the dynamical laws, the solutions still cannot be extended in time to make reasonable physical sense. For elastodynamics non-essential shock type singularities are well-known, but there may be essential singularities as well. The analogy with celestial mechanics is clear: some singularities, like collisions, are inessential and solutions can be continued beyond them (by various regularization devices, including _ad hoc_ postulates such as conservation of linear and angular momentum), while others, in which infinite velocities in a finite time develop, seem essential. For the Euler equations for an ideal fluid in three dimensions with periodic boundary conditions, essential singularities seem likely. (See Chorin, Hughes, McCracken and Marsden [46] for a discussion.)

Definition Let S be a topological space. A _continuous flow_ (resp. semi-flow) on S is a continuous map $F: D \subset S \times \mathbb{R} \longrightarrow S$ (resp. $F: D \subset S \times \mathbb{R}^{+} \longrightarrow S$) where D is open, such that

(i) $S \times \{0\} \subset D$ and $F(x,0) = x$,

(ii) if $(x,t) \in D$, then $(x,t+s) \in D$ iff $(F(x,t),s) \in D$ and in this case $F(x,t+s) = F(F(x,t),s)$.

For general properties of flows and semi-flows such as separate continuity in t and x implying continuity etc., see Chernoff and Marsden [39] and Marsden and McCracken [185]. We shall usually write $F_t(x) = F(x,t)$, so (i) and (ii) respectively read $F_0 = \mathrm{id}$, $F_{t+s} = F_s \circ F_t$ (where defined).

For nonlinear elastodynamics, the proof that the equations define a continuous flow is not so simple, especially the continuity in Sobolev spaces

H^s with the lowest possible s. See Section 10 and Hughes, Kato and Marsden [129] and Kato [142].

The "equations" generating the semi-flow F_t constitute the generator, defined in a way similar to the linear case.

Definition Let F_t be a semi-flow on a set D in a Banach space (or manifold) X. The X-infinitesimal generator of F_t is the map

$$G(x) = \frac{d}{dt+} F_t(x) \big|_{t=0},$$

taken in X on the domain D(G) for which the derivative exists.

We shall next give a result of Dorroh-Marsden [70] which is applicable to quasi-linear hyperbolic systems. The theorems concern the differentiability of F_t for t fixed. Such results are important for understanding the sense in which F_t is a canonical transformation. For theorems applicable to parabolic systems, see Marsden and McCracken [185]. Before commencing, we remark on the delicacy of the results. Indeed, the example $u_t + uu_x = 0$ shows that $F_t : H^s \to H^s$, $s > \frac{3}{2}$ is continuous, but not Hölder continuous for any positive exponent. See Kato [139] for the proof. On the other hand, $F_t : H^s \to H^{s-1}$ is differentiable and its derivative extends to a bounded operator on H^{s-1}.

We shall need to define the relevant notions of differentiability for G and for F_t in order to state the result. First, for the generator.

Definition Let X and Y be Banach spaces with $Y \subset X$ continuously and densely included. Let $U \subset Y$ be open and $F : U \to X$ be a given mapping. We say f is α-differentiable if, for each $x \in U$, there is a bounded linear operator $Df(x) : Y \to X$ such that

$$\| f(x+h) - f(x) - Df(x)\cdot h \|_X / \| h \|_X \to 0$$

as $\|h\|_Y \longrightarrow 0$. If f is α-differentiable and $x \mapsto Df(x) \in B(Y,X)$ is norm continuous, we call f C^1 α-differentiable. (Notice that this is stronger than C^1 in the Frechet sense.) If f is α-differentiable and $\|f(x+h) - f(x) - Df(x)\cdot h\|_X / \|h\|_X$ is uniformly bounded for x and $x+h$ in some Y neighborhood of each point, we say that f is locally uniformly α-differentiable.

Most concrete examples can be checked using the following:

Proposition Suppose $f: U \subset Y \longrightarrow X$ is of class C^2 and Y- locally in x,

$$\|D^2f(x)(h,h)\|_X / \|h\|_Y \|h\|_X$$

is bounded. Then f is locally uniformly C^1 α-differentiable.

This follows easily from the identity

$$f(x+h) - f(x) - Df(x)\cdot h = \int_0^1\int_0^1 D^2f(x+sth)(h,h)dsdt$$

Next we turn to the appropriate notion for the flow.

Definition A map $g: U \subset Y \longrightarrow X$ is called β-differentiable if it is α-differentiable and $Dg(x)$, for each $x \in U$, extends to a bounded operator of X to X.

The nice thing about β-differentiable maps is that they obey a chain rule. For example, if $g_1: Y \longrightarrow Y$, $g_2: Y \longrightarrow Y$ and each are β-differentiable (as maps of Y to X) and are continuous from Y to Y, then $g_2 \circ g_1$ is β-differentiable with, of course,

$$D(g_2 \circ g_1)(x) = Dg_2(g_1(x)) \circ Dg_1(x)$$

The proof of this fact is routine. In particular, one can apply the chain rule to $F_t \circ F_s = F_{t+s}$ if each F_t is β-differentiable. Differentiating this in s at $s=0$ gives

$$DF_t(x)\cdot G(x) = G(F_t(x)),$$

the flow invariance of the generator. For completeness we formally state this result and give another, direct proof as it illustrates just how β-differentiability is needed. (The proof of the chain rule mentioned above uses similar ideas.)

<u>Proposition</u> Suppose $Y \subset X$ and F_t is a continuous semiflow on an open set $D \subset Y$ with X-infinitesimal generator $G : D \longrightarrow X$. Suppose that for each $x \in D$ and $0 \leqslant t < T_x$, F_t is β-differentiable at x. Then

$$G(F_t(x)) = DF_t(x)\cdot G(x)$$

for $x \in D$ and $0 \leqslant t < T_x$.

<u>Proof</u>. $\frac{1}{h}[F_h(F_t(x) - F_t(x)] - DF_t(x)\cdot\frac{1}{h}(F_h(x) - x)$

$$= \frac{1}{h}(F_t(F_h(x)) - F_t(x)) - DF_t(x)\cdot\frac{1}{h}(F_h(x) - x).$$

In Y norm $F_h(x) - x \longrightarrow 0$ as $h \longrightarrow 0$ while in X norm $\frac{1}{h}(F_h(x) - x) \longrightarrow G(x)$. Thus by definition of the β-derivative, the displayed expression tends to zero. It also converges, since $DF_t(x) \in B(X,X)$, to $G(F_t(x)) - DF_t(x)\cdot G(x)$, so our claim holds. □

One can also use the chain rule to prove a uniqueness theorem for integral curves of generators of semi-flows of β-differentiable maps by much the same method as we used for the linear case in §5, so the details may be omitted. Thus in this context, which includes elastodynamics one can say that <u>existence of semiflows implies uniqueness of solutions of the Cauchy problem</u>. (In Chernoff-Marsden [39] this was proved, by the same method, but under hypotheses which include quasi-linear parabolic and semi-linear hyperbolic equations, but which exclude the quasi-linear hyperbolic case. The present version includes all of these.)

For the following theorem we assume these hypotheses: $Y \subset X$ is continuously and densely included and F_t is a continuous semi-flow on an open subset $D \subset Y$ and the X-infinitesimal generator G of F_t has domain D. Also, we assume:

(H_1) $G : D \subset Y \longrightarrow X$ is locally uniformly C^1 α-differentiable,

(H_2) for $x \in D$, let T_x be the lifetime of x, i.e.,

$\{t \in \mathbb{R} \mid F_t^{\cdot}(x)$ is defined$\}$. Assume there is a strongly continuous linear evolution system $\{U^x(t,s) : 0 \leqslant s \leqslant t < T_x\}$ in X whose X-infinitesimal generator is an extension of $\{DG(F_s x) \in B(Y,X);$ $0 \leqslant s < T_x\}$; i.e., if $y \in Y$,

$$\frac{\partial}{\partial t^+} U^x(t,s)\cdot y \big|_{t=s} = DG(F_s(x))\cdot y.$$

For the theory of linear evolution systems (i.e., time dependent linear semigroups), we refer to Kato [137,138]. The results relevant for the present discussion are reproduced in Marsden and McCracken [185]. The following is the principal result.

Theorem Under the above hypotheses, F_{t-s} is β-differentiable at $F_s(x)$ and in fact,

$$DF_{t-s}(F_s(x)) = U^x(t,s).$$

Proof. Define $\phi(x,y)$ by

$$G(x) - G(y) - DG(y)\cdot(x - y) + \|x - y\|_X \, \phi(x,y)$$

(or zero if $x = y$) and notice that by local uniformity, $\|\phi(x,y)\|_X$ is uniformly bounded if x and y are Y-close. By joint continuity of $F_t(x)$, for $0 < T < T^X$, $\|\phi(F_s y, F_s x)\|_X$ is bounded for $0 \leqslant s \leqslant T$ if $\|x - y\|_Y$ is sufficiently small.

As in the linear case, we have the equation

$$\frac{d}{ds} F_s(x) = G(F_s(x)), \qquad 0 \leqslant s \leqslant T, \qquad x \in D.$$

Let

$$w(s) = F_s(y) - F_s(x)$$

so that

$$\begin{aligned} w'(s) &= G(F_s(y)) - G(F_s(x)) \\ &= DG(F_s(x))w(s) + \|w(s)\|_X \, \phi(F_s y, F_s x). \end{aligned}$$

Since $DG(F_s x)\cdot w(s)$ is continuous in s with values in X, and writing $U = U^X$, we have the backwards differential equation:

$$\frac{\partial}{\partial s} U(t,s)w(s) = U(t,s)w'(s) - U(t,s)DG(F_s(x))\cdot w(s).$$

Hence

$$w(t) = U(t,0)(y - x) + \int_0^t U(t,s) \|w(s)\|_X \phi(F_s(y), F_s(x))\cdot ds.$$

Let $\|U(t,s)\|_{X,X} \leqslant M_1$ and $\|\phi(F_s(y), F_s(x))\|_X \leqslant M_2$, $0 \leqslant s \leqslant t \leqslant T$. Thus, by Gronwall's inequality,

$$\|w(t)\|_X \leqslant M_1 e^{M_1 M_2 T} \|y - x\|_X = M_3 \|y - x\|_X.$$

In other words,

$$\|F_t(y) - F_t(x) - U(t,0)(y-x)\|_X / \|y-x\|_X \leqslant M_1 M_3 \int_0^t \|\phi(F_s(y), F_s(x)\|_X ds.$$

From the bounded convergence theorem†, we conclude that F_t is β-differentiable at x and $DF_t(x) = U(t,0)$. This is the theorem for $s = 0$. The general result follows by making a translation. □

A technical problem one should note is that in general one does not have the forward differential equation that one might expect from semigroup theory (when $U(t,s) = U(t-s)$); i.e.,

$$\frac{\partial}{\partial t} U(t,s)y = DG(F_t(x))\cdot U(t,s)y$$

need not hold. The trouble is that $U(t,s)$ need not map the domain of $DG(F_s(x))$ into that of $DG(F_t(x))$. On the other hand, we do have the backwards equation:

$$\frac{\partial}{\partial s} U(t,s)y = - U(t,s)DG(F_s(x))\cdot y$$

as was used in the previous proof. In order for the forward differential equation to be satisfied we need to assume that the evolution system $U(t,s)$ is Y-stable, i.e., $U(t,s)Y \subset Y$. In his papers, Kato gives a number of conditions guaranteeing this. It is satisfied for quasi-linear hyperbolic systems of second order and for quasi-linear symmetric hyperbolic systems.

We shall now show how these results apply to Hamiltonian systems. We shall temporarily assume that the spaces are linear for simplicity, but this is not essential and will be abandoned shortly. Assume $Y \subset X$ is as above and F_t is a continuous (semi-) flow on Y with X-infinitesimal generator

---

† $\phi(F_s(y), F_s(x))$ is strongly measurable in s since $\phi(x,y)$ is continuous for $x \neq y$.

G defined on an open set $D \subset Y$. Let $\omega : X \times X \longrightarrow \mathbb{R}$ be a weakly non-degenerate skew symmetric bilinear form. As in the linear case, we say G is Hamiltonian if there is a C^1 function $H : D \longrightarrow \mathbb{R}$ such that

$$\omega(G(x),y) = dH(x)\cdot y$$

for all $x,y \in Y$. If G is C^1, this is equivalent to saying that $DG(x)$ is ω-skew, by the Poincaré lemma or directly. Indeed if G is C^1 and Hamiltonian, H is C^2, since $dH(x) = \omega(G(x),\cdot)$. Differentiating gives $\omega(DG(x)\cdot y_1, y_2) = d^2H(x)\cdot(y_1,y_2)$. Since $d^2H(x)$ is symmetric in y_1, y_2, $DG(x)$ must be ω-skew. Conversely, if $DG(x)$ is ω-skew we can choose

$$H(x) = \int_0^1 \omega(G(tx),x)dt + \text{Constant}$$

as is easily checked. (This is analogous to the criteria for an operator to be potential; see Vainberg [250] and Hughes and Marsden [130]).

Proposition In addition to the hypotheses of the previous theorem, assume the evolution system $U(t,s)$ is Y-regular. If G is Hamiltonian, then $F_t^*\omega = \omega$, i.e.,

$$\omega(DF_t(x)\cdot y_1,\ DF_t(x)\cdot y_2) = \omega(y_1,y_2)$$

for $x \in D$ and $y_1, y_2 \in X$.

Proof. For $y_1, y_2 \in Y$, we have

$$\frac{d}{dt}\,\omega(DF_t(x)\cdot y_1, DF_t(x)\cdot y_2) = \frac{d}{dt}\,\omega(U(t,0)\cdot y_1, U(t,0)\cdot y_2)$$

$$= \omega(DG(F_t(x))\cdot U(t,0)\cdot y_1, U(t,0)\cdot y_2) + \omega(U(t,0)\cdot y_1, DG(F_t(x))\cdot U(t,0)\cdot y_2)$$

which vanishes because DG is ω-skew. This proves the given result in case $y_1, y_2 \in Y$. Since Y is dense in X, the conclusion for $y_1, y_2 \in X$ follows by continuity. □

In this result we took Y and X to be linear spaces and ω a constant bilinear form only for convenience. The result also holds on general (weak) symplectic manifolds; the proof is only slightly more complicated. We can conclude from all this that a satisfactory theory of canonical transformations can be built around the concept of β-differentiable maps.

For conservation laws one can do somewhat better. For example, when shock waves for quasi-linear hyperbolic systems develop, the result below shows that the solutions cannot be continued into spaces in which (i) the dynamics still defines a C^0 semi-flow, (ii) in which the generator is Hamiltonian and (iii) for which energy is not conserved when shocks form. This is because we shall prove a general conservation theorem (including conservation of energy) under the rather weak assumptions (i) and (ii) just stated.

Indeed, when shocks form it is well-known that uniqueness fails unless entropy conditions are imposed. The conservation law below may be interpreted by saying that in elastodynamics a lack of well defined dynamics from the Hamiltonian structure alone must accompany a lack of energy conservation[†].

As we shall see shortly and in Section 10, the Hamiltonian structure makes natural sense for weak solutions. It is, perhaps, an interesting problem to make sense of the semi-flow defined by shock solutions as a dissipative Hamiltonian system; cf. Quinn [214], Crandall [59] and Dafermos [64]; the latter reference applies to one-dimensional elasticity. We turn to the

---

[†]In thermo-elasticity, where energy is probably conserved through the shock, this remark is irrelevant.

conservation theorem under discussion. We let $Y \subset X$ as above and let F_t be a C^0 semi-flow on an open set $D \subset Y$ with generator G. Assume $G : D \longrightarrow X$ is continuous and is Hamiltonian with energy H. Let $G_K : D \longrightarrow X$ be another continuous operator which is Hamiltonian with energy a C^1 map $K : D \longrightarrow \mathbb{R}$. Let the Poisson Bracket be given by

$$\{K,H\}(x) = \omega(G_K(x),G(x)) \qquad \text{for} \quad x \in D$$

Theorem (Chernoff-Marsden). Under the assumptions just stated, for $x \in D$, $K(F_t(x))$ is t-differentiable with derivative

$$\frac{d}{dt} K(F_t(x)) = \{K,H\}(F_t(x))$$

for $0 \leqslant t < T_x$, the lifetime of x. In particular if $\{K,H\} = 0$, then $K(F_t(x)) = K(x)$ and so, taking $K = H$, it follows that $H(F_t(x)) = H(x)$.

Remark The tricky point is that $K(F_t(x))$ is not obviously t-differentiable under these hypotheses; the chain rule does not apply since $t \longmapsto F_t(x) \in D$ is not differentiable in the Y-topology.

Proof. It suffices to prove that for each $x_0 \in D$,

$$\frac{d}{dt} K(F_t(x_0))\Big|_{t=0} = \{K,H\}(x_0)$$

by the semi-flow property of F_t. Also, we can take $x_0 = 0$ by translating the semi-flow.

As remarked earlier, we can relate K to G_K by

$$K(x) = K(0) + \int_0^1 \omega(G_K(\tau x),x)d\tau.$$

Let $x_t = F_t(0)$. Then

$$\frac{1}{t}\{K(x_t) - K(0)\} = \int_0^1 \omega(G_K(\tau x_t), \frac{x_t}{t})d\tau.$$

Now as $t \to 0$, $x_t \to 0$ in D and as G_K is continuous, $G_K(\tau x_t) \to G_K(0)$ uniformly for $0 \leqslant \tau \leqslant 1$. Also $\frac{x_t}{t} = \frac{x_t - x_0}{t} \to G(0)$ as $t \to 0$ since G is the generator of F_t. Thus, we can pass to the limit under the integral sign to obtain

$$\underset{t\to 0+}{\text{limit}} \frac{1}{t}\{K(x_t) - K(0)\} = \int_0^1 \omega(G_K(0),G(0))d\tau = \omega(G_K(0),G(0)) = \{K,H\}(0). \qquad \square$$

Once this basic result is established one can go on to more sophisticated geometric set-ups, such as general conserved quantities (or momentum mappings) associated with Lie group actions, as in Abraham and Marsden [1], Chapter 4.

For applications in the next section it will be necessary to use a number of results concerning Lagrangian systems. These applications also must allow X and Y to be manifolds. (This changes little of what we have already done.) For example, the basic configuration space for the place problem of elastodynamics is a nonlinear space $\mathscr{K}$ of maps; it is not even an open set in a linear space, as we shall see. For the traction problem it is an open set in a linear space, but this is lost if constraints such as incompressibility are imposed. Details are given in the following section.

The material which follows is standard and summarizes what we need for the readers' convenience. For additional details and references we refer to Abraham and Marsden [1] and Chernoff and Marsden [39].

Let M be a manifold modelled on a Banach space X. Let T^*M be its cotangent bundle, and $\pi^* : T^*M \to M$ the projection. The <u>canonical one form</u> θ on T^*M is the one form defined by

$$\theta(\alpha)\cdot w = \alpha\cdot T\pi^*(w),$$

where $\alpha \in T_x^*M$, $x \in M$ and $w \in T_\alpha(T^*M)$. In a local coordinate chart $U \subset X$, this reads:

$$\theta(x,\alpha)\cdot(y,\beta) = -\alpha(y),$$

where $(x,\alpha) \in U \times X^*$, $(y,\beta) \in X \times X^*$. If M is finite dimensional, this formula reads

$$\theta = \Sigma\, p_i dq^i$$

where $q^1, \ldots, q^n$, $p_1, \ldots, p_n$ are standard coordinates for T^*M.

The canonical two form is defined by $\omega = -d\theta$, where d is the exterior derivative. Using the local formula for d one computes that in a local coordinate chart $U \subset X$ for T^*M,

$$\omega(x,\alpha)\cdot((y_1,\alpha_1),(y_2,\alpha_2)) = \alpha_2(y_1) - \alpha_1(y_2)$$

which coincides with the canonical symplectic form in the linear case discussed in Section 6. In the finite dimensional case, the formula for ω is

$$\omega = \Sigma\, dq^i \wedge dp_i .$$

The canonical two form ω on T^*M is a weak symplectic form since $d\omega = 0$ (because $d^2 = 0$) and since ω on each tangent space is a weak symplectic form in the linear sense.

Next we consider symplectic forms induced by metrics. If $\langle\, ,\rangle_x$ is a weak Riemannian (or pseudo-Riemannian) metric on M, we have a smooth map $\phi : TM \longrightarrow T^*M$ defined by $\phi(v_x)w_x = \langle v_x, w_x \rangle_x$, where $x \in M$ and $v_x, w_x \in T_xM$. If $\langle\, ,\rangle$ is a (strong) Riemannian metric it follows from the implicit function theorem that ϕ is a diffeomorphism of TM onto T^*M but this is not the situation in most examples. In any case, set $\Omega = \phi^*(\omega)$, the pull-back of ω by ϕ where ω is the canonical form on T^*M. Clearly Ω is exact since $\Omega = d(\phi^*(\theta))$.

Using the definition of pull-back we can readily verify the following.

<u>Proposition</u> (a) If $\langle\ ,\rangle_x$ is a weak metric, then Ω is a weak symplectic form. In a chart $U \subset X$ for M we have

$$\Omega(x,y)((x_1,y_1),(x_2,y_2)) =$$
$$= D_x\langle y,x_1\rangle_x x_2 - D_x\langle y,x_2\rangle_x x_1 + \langle y_2,x_1\rangle_x - \langle y_1,x_2\rangle_x$$

where D_x denotes the Fréchet derivative with respect to x.

(b) If $\langle\ \rangle_x$ is a strong metric and M is modelled on a reflexive space, then Ω is a strong symplectic form.

(c) $\Omega = d\Theta$ where, locally, $\Theta(x,e)(e_1,e_2) = -\langle e,e_1\rangle_x$.

<u>Note</u> In the finite dimensional case, the formula for Ω becomes

$$\Omega = \Sigma g_{ij} dq^i \wedge d\dot{q}^j + \Sigma \frac{\partial g_{ij}}{\partial q^k} \dot{q}^i dq^j \wedge dq^k$$

where $q^1, \ldots, q^n, \dot{q}^1, \ldots, \dot{q}^n$ are coordinates for TM.

Generalizing the case of $Y \subset X$, two Banach spaces with the inclusion dense and continuous, let us call a <u>manifold domain</u> of P, a subset $D \subset P$ such that D has its own manifold structure for which the inclusion $i : D \longrightarrow P$ is C^∞ and such that its tangent $Ti : TD \longrightarrow TP$ is also injective.

For the nonlinear case we shall refer alternatively to generators as vector fields. If (P,ω) is a weak symplectic manifold, the condition that a vector field $G : D \longrightarrow TP$ (with manifold domain $D \subset TP$) be Hamiltonian with energy H is that

$$i_G\omega = dH,$$

i.e., for $x \in P$, and $v \in T_x P$,

$$\omega_x(G(x), v) = dH(x) \cdot v.$$

The reader will note that this is a generalization of the definition in the linear case. Also note that locally, G is Hamiltonian iff $d(i_G \omega) = 0$; in a chart in which ω is constant[†], this says that $DG(x)$, a linear operator, is ω-skew, just as in the linear case.

We shall write $G = G_H$ because usually in examples H is given and then one constructs the Hamiltonian vector field G_H. Of course in the finite dimensional case, where $\omega = \Sigma\, dq^i \wedge dp_i$, the generator takes the familiar form of Hamilton's equations:

$$G_H = \left(\frac{\partial H}{\partial p_i}, -\frac{\partial H}{\partial q^i} \right).$$

Because ω is only weak, given $H : D \longrightarrow \mathbb{R}$, G_H need not exist. Also, even if H is smooth on all of P, G_H will in general be defined only on a certain subset D of P, but where it is defined, it is unique. The linear wave equation discussed in Section 6 is a case in point. Notice, however, that even if H is only defined on D, for $x \in D$, $dH(x)$ must extend to a bounded linear functional on $T_x P$ because of the defining relation between G and H. Thus in the obvious sense H is β-differentiable.

The reader is invited to extend the proofs given above on conservation of energy and the symplectic nature of the flow to the case in which the phase space is not necessarily linear and ω is not necessarily constant.

[†]Such charts always exist if ω is strongly nondegenerate, (Darboux's Theorem). In general they need not (Marsden [180]) although they do under reasonable conditions which are fulfilled in all examples of interest (Tromba [241]).

For elastodynamics we shall be interested in Lagrangian systems, so we now recall some of their essential features.

Let M be a manifold modelled on a Banach space X_1, and let $Q \subset M$ be a manifold domain. Consider the following subset of the tangent bundle TM:

$$P = \bigcup_{m \in Q} T_m M.$$

We call P the restriction of TM <u>to</u> Q and write $P = TM|Q$. If Y_1 is the model space for Q, we can endow P with a manifold domain structure by giving it the local product structure $X = Y_1 \times X_1$.

By a <u>Lagrangian</u> on $P \subset TM$ we mean a smooth function $L : P \longrightarrow \mathbb{R}$. In particular, for each $m \in Q$, L restricts P to a smooth function on $T_m M$. Thus we can form the fiber derivative $FL : P \longrightarrow T^*M$; explicitly, if $v, w \in T_m M$, we have the formula

$$FL(v)\cdot w = \frac{d}{dt} L(v + tw)\Big|_{t=0}.$$

Define $\omega_L = (FL)^*\omega$, a two form on P where ω is the canonical two form on T^*M. Thus $\omega_L = -d\Theta_L$ where $\Theta_L = (FL)^*\Theta$, Θ being the canonical one form on T^*M. Locally, one has the formula

$$\omega_L(x,v)\ ((x_1,v_1),(x_2,v_2)) = D_1(D_2L(x,v)\cdot x_1)\cdot x_2 - D_1(D_2L(x,v)\cdot x_2)\cdot x_1$$
$$+ D_2D_2L(x,v)\cdot v_2\cdot x_1 - D_2D_2L(x,v)\cdot v_1 x_2 .$$

We will call L <u>regular</u> if ω_L is weakly non-degenerate; i.e., if $D_2D_2L(x,v)$ is weakly non-degenerate. This assumption will be made here.

The <u>action</u> A and <u>energy</u> E are defined by

$$A(v) = FL(v)\cdot v,$$

$$E = A - L.$$

Let D be the subset of P consisting of all points v such that $G_E(v) \in TP$ is defined. Thus we regard G_E as a vector field on P with domain D. We call G_E the <u>Lagrangian vector field</u> determined by L.

The classical Lagrange equations are second-order equations. There is a general notion of second-order fields on a tangent bundle TM, and we can extend this notion to the case of vector fields defined on $P \subset TM$. Indeed, the projection $\pi : TM \longrightarrow M$ restricts to a map from P to Q; we say that G_E is of <u>second order</u> iff $T\pi(G_E(v)) = v$ for each $v \in D$.

The following is proved as in finite dimensions (see Abraham and Marsden [1]).

<u>Theorem</u> Let L be a regular Lagrangian, with associated vector field G_E defined on $D \subset P$ as discussed above. Conclusions:

(i) $D \subset TQ$;

(ii) X_E is a second-order vector field;

(iii) In local coordinates, a point (x,v) of TQ belongs to D if and only if

$$\phi(x,v) = D_1L(x,v) - D_2(D_1L(x,v)\cdot v)$$

lies in the range of $D_2D_2L(x,v)$, regarded as a map from X_1 into Y_1^*. If this condition is met, we have the formula

$$G_E(x,v) = (v, [D_2D_2L(x,v)]^{-1}\cdot\phi(x,v));$$

(iv) A curve $c(t)$ in Q is a base integral curve of G_E if and only if <u>Lagrange's equations</u> hold:

$$\frac{d}{dt}D_2L(c(t), c'(t)) = D_1L(c(t), c'(t)).$$

Remarks 1. As we shall see, this form of Lagrange's equations for field theories is a statement of the weak form of the equations.

2. By a base integral curve we mean the projection to Q of an integral curve of G_E. Explicitly, let $\tilde{c}(t) \in D$ be an integral curve of G_E. (This means that $\tilde{c}'(t) = G_E(\tilde{c}(t))$, where $\tilde{c}'(t)$ is computed relative to the manifold structure of P.) Define $c(t) = \pi(\tilde{c}(t))$, a curve in Q. Because G_E is a second-order vector field, we have $c'(t) = \tilde{c}(t)$, where the derivative is computed relative to the manifold structure of M.

3. Because E is a smooth function on P, it was not necessary to introduce a manifold structure on $\dot{N}$ in order to construct G_E. Actually, in many instances D will have a manifold structure, akin to the graph topology in the linear case.

To illustrate these constructions, consider again the wave equation. We start with $M = X_1 = L^2(\mathbb{R}^n)$ and the Lagrangian

$$L(\phi,\dot{\phi}) = \tfrac{1}{2}\langle \dot{\phi},\dot{\phi}\rangle - \tfrac{1}{2}\langle \nabla\phi,\nabla\phi\rangle$$

defined on the space $P = H^1 \times L^2$. Note that P is just the restriction of $L^2 \times L^2 = TL^2$ to $Q = H^1 \subset L^2$. The fiber derivative of L is the map from $H^1 \times L^2$ to $L^2 \times (L^2)^* = T^*L^2$ given by the formula

$$FL(\phi,\dot{\phi}) = (\phi,\langle \dot{\phi},\cdot\rangle).$$

Hence we have

$$\omega_L((\phi_1,\dot{\phi}_1), (\phi_2,\dot{\phi}_2)) = \{\langle \dot{\phi}_2,\phi_1\rangle - \langle \dot{\phi}_1,\phi_2\rangle\};$$

This is the same symplectic form we used in Section 6 for the wave equation. We know that ω_L is weakly non-degenerate on P. By a straightforward computation we find the energy to be

$$E(\phi,\dot\phi) = \tfrac{1}{2}\langle \dot\phi,\dot\phi\rangle + \tfrac{1}{2}\langle \nabla\phi,\nabla\phi\rangle,$$

whence we have

$$dE(\phi,\dot\phi)\cdot(\psi,\dot\psi) = \langle \dot\phi,\dot\psi\rangle + \langle \nabla\phi,\nabla\psi\rangle.$$

Next, let us use the above theorem to determine the domain D of G_E. We know that D is contained in $Q = H^1 \times H^1$. We must consider $\alpha(\psi) = -\langle \nabla\phi,\nabla\psi\rangle$ as a function of ψ; when does α lie in the range of $D_2D_2L(\phi,\dot\phi) = \langle \dot\phi,\cdot\rangle$? This is so if and only if $\phi \in H^2$, in which case $\alpha(\psi) = \langle \Delta\phi,\psi\rangle$ and so

$$G_E(\phi,\dot\phi) = (\dot\phi,\Delta\phi)$$

on the domain $D = H^2 \times H^1$. This is then in accord with our treatment of the wave equation in Section 6.

We now generalize the above example. Suppose that M is a (weak) Riemannian manifold with metric $\langle\cdot,\cdot\rangle$. Let $V : Q \subset M \longrightarrow \mathbb{R}$ be a smooth function on a manifold domain Q. Define a Lagrangian by

$$L(v_q) = \tfrac{1}{2}\langle v_q, v_q\rangle - V(q), \quad v_q \in T_qM, \quad q \in Q.$$

L is a smooth function on $P = TM|Q$. Moreover, the fiber derivative of L does not involve V; in fact it is just the map from TM to T^*M determined by the metric $\langle\cdot,\cdot\rangle$. Hence ω_L is the weak symplectic form induced by the metric.

We say that the weak Riemannian metric has a <u>smooth spray</u> provided there is a smooth Lagrangian vector field $G_K : TM \longrightarrow T(TM)$ associated to the kinetic energy function $K(v) = \tfrac{1}{2}\langle v,v\rangle$ on TM. (It is customary to give a geometric definition of sprays in terms of connections. Our approach leads to the same thing.) In local coordinates, the condition for existence of a spray is that there be a smooth map $G_2(x,v)$ satisfying the relation

$$\langle Y_2(x,v),v_1\rangle_x = \tfrac{1}{2}D_x\langle v,v\rangle_x\cdot v_1 - D_x\langle v,v_1\rangle_x\cdot v.$$

The spray G_K is then given by the local formula

$$G_K(x,v) = (x,G_2(x,v)).$$

It is easy to see that $G_2(x,v)$ depends quadratically on v; this is a characteristic property of sprays.

Even though the metric $\langle\ ,\ \rangle$ is weak, it may nevertheless posess a smooth spray. This occurs in elastodynamics and fluid mechanics as we shall see in the next section. It also occurs in a number of other situations as well (see Fischer and Marsden [84], Choquet-Bruhat and Marsden [45], Ebin [75] and Tromba [242]).

If $q \in Q$, we say that grad $V(q)$ exists provided there is a vector $u \in T_qM$ such that, for all $v \in T_qQ$,

$$dV(q)\cdot v = \langle u,v\rangle.$$

We write $u = \operatorname{grad} V(q)$ in this case, and let $D_0 = \{q \subset Q : \operatorname{grad} V(q) \text{ exists}\}$, the domain of grad V.

Using these definitions, we generalize a familiar theorem of Lagrangian mechanics to the infinite dimensional case by a straightforward calculation.

Theorem Let $\langle\cdot,\cdot\rangle$ be a weak Riemannian metric on M which has a smooth spray. Let $V : Q \subset M \longrightarrow \mathbb{R}$ be a smooth function with manifold domain Q, and define L on $P = TM|Q$ as above: $L(v_q) = K(v_q) - V(q)$. Assume that the spray G_K exists and maps P into TP.

The domain D of G_E, the associated Lagrangian vector field, is $D = TQ|D_0$ where D_0 is the domain of grad V. We have the formula

$$G_E(v_q) = G_K(v_q) - [\operatorname{grad} V(q)]^{\ell}_{v_q}, \qquad v_q \in D,$$

where $[\ \ ]^{\ell}_{v_q}$ denotes the canonical injection ("vertical lift") of T_qM into $T_{v_q}(TM)$ given by

$$[u_q]^{\ell}_{v_q} = \frac{d}{dt}(v_q + tu_q)\big|_{t=0}.$$

In the next section we will briefly discuss some problems connected with incompressible elasticity. To do so we shall need some general facts about constrained systems in the context of the present discussion. A crucial fact we wish to emphasise is that in many cases the property of having a smooth spray is not destroyed by the presence of constraints. For the case of an ideal fluid with the constraint of incompressibility this fact was discovered by Ebin and Marsden [77]. It has considerable technical utility for analytical purposes as well as being of interest in its own right. In the next section we shall explain how this same result can be used in elasticity.

<u>Theorem</u> Let M be a weak Riemannian manifold possessing a smooth spray $S : TM \longrightarrow T^2M$. Let N be a submanifold of M. Suppose that, for each $n \in N$, there is an orthogonal decomposition† $T_nM = T_nN \oplus C_n$. Using this decomposition, define a projection $P : TM|N \longrightarrow TN$. Assume P is smooth. Then the restriction to N of the Riemannian metric has a smooth spray, given by

$$S_N = TP \circ S \quad \text{at points of} \quad TN.$$

The verification of this theorem is straightforward. The difference

---

† The existence of such decompositions is automatic for strong metrics. For weak metrics it usually relies on the Fredholm alternative and elliptic theory.

$S_N - S$ is vertical over N, and so it may be identified with a vector field on $N \subset M$. This vector field is orthogonal to N; intuitively, it gives the "forces of constraint" which ensure that particle trajectories remain in N. For incompressible fluid dynamics or elasticity, this force of constraint is, of course, the pressure gradient.

In addition to the above considerations, we also will wish to regard elasticity as a Lagrangian field theory in the classical sense. We can then make use of the Noether theorem on conservation laws, etc. The remainder of this section reviews this topic. (See, for example, Hermann [121] and references therein.)

Most classical field theories work in the context of vector or tensor valued fields. This means that one deals with sections of a <u>vector</u> bundle. Examples like elasticity show that one should in fact be dealing with <u>fiber</u> bundles. (As we already saw in the first section the fields ϕ in elasticity do not transform like vectors but rather like point mappings.)

Let $\pi : E \longrightarrow M$ be a fiber bundle and let $J^1(E)$ be its first jet bundle (see, e.g., Palais [207]). By definition, its fiber over a point $x \in M$ can be written

$$J^1(E)_x = \{(\phi_x, \delta_x) \mid \phi_x \in E_x = \pi^{-1}(x) \text{ and } \delta_x \text{ is a linear map of } T_xM \text{ to } T_{\phi_x}E_x\}.$$

We shall assume that M is endowed with a volume element μ and that E is equipped with Riemannian structure on each fiber as well as a connection. If ϕ is a section of E, then its first jet is the section of $J^1(E)$ given by

$$j(\phi) = \phi \oplus D\phi$$

where $D\phi$ is the covariant derivative of ϕ.

To incorporate the velocities $\dot{\phi}$ we need to take a variation of the fields ϕ. Let us define a new bundle $\dot{E}$ over M by giving its fiber over $x \in M$:

$$\dot{E}_x = \{(\phi_x, \dot{\phi}_x) \mid \phi_x \in E_x, \ \dot{\phi}_x \in T_{\phi_x} E_x\}.$$

Define the bundle F over M whose fiber over x consists of triples $\dot{\phi}_x$, ϕ_x, δ_x

$$F_x = \{(\phi_x, \dot{\phi}_x, \delta_x) \mid (\phi_x, \dot{\phi}_x) \in \dot{E}_x, \ (\phi_x, \delta_x) \in J^1(E)_x\},$$

(A "diagonal" in $J^1(E) \oplus \dot{E}$).

A Lagrangian <u>density</u> is a map

$$\mathcal{L} : F \longrightarrow \mathbb{R}$$

and we shall, with abuse of notation, write

$$\mathcal{L}(x, \phi, \dot{\phi}, D\phi).$$

(For much of what follows, x is irrelevant and is suppressed.)

Let $\mathcal{H}$ denote the space of sections of the bundle E of an appropriate differentiability class. (For elasticity, boundary conditions of place are to be imposed on $\mathcal{H}$ as well; for the traction problem, see below.) Then $T\mathcal{H}$ can be identified with the space of sections of the bundle $\dot{E}$. (For $\mathcal{H}$ the space of configurations in elasticity, the bundle is $M \times N \longrightarrow M$ with a configuration $\phi : M \longrightarrow N$ identified with the section $X \longmapsto (X, \phi(X))$. Then the description of $T\mathcal{H}$ here coincides with that of Section 4.)

Writing a section of $\dot{E}$, i.e., an element of $T\mathcal{H}$, as $(\phi, \dot{\phi})$ we define the Lagrangian $L(\phi, \dot{\phi})$ by

$$L : T\mathscr{H} \longrightarrow \mathbb{R}$$

$$L(\phi,\dot{\phi}) = \int_M \mathcal{L}(x,\phi(x),\dot{\phi}(x),D\phi(x))d\mu(x).$$

Notice that $\mathcal{L}$ can depend explicitly on the base point $x \in M$, because $\mathcal{L}$ is a map on $E \oplus J^1(E)$, and the fibers depend on the base point. In general relativity the metric is one of the fields, so μ depends on the fields. In this case it is better to regard $\mathcal{L}$ as a density than a scalar. In our discussions in Section 2, the metric g was included as a variable in $\mathcal{L}$ but only as a parameter; it was not a field variable. In relativistic elasticity, g will be a field variable; our formalism developed in Sections 2 and 3 makes coupling with gravity fairly natural.

We choose for the configuration space $Q = \mathscr{H}$ a suitable Sobolev class of sections of E. With the appropriate choice one can prove that $L(\phi,\dot{\phi})$ is a smooth function of $(\phi,\dot{\phi})$ by using the composition results of Section 4. Then one can easily establish formulae like the following (using the obvious notation):

$$DL(\phi,\dot{\phi})\cdot(h,\dot{h}) = D_{\dot{\phi}}L(\phi,\dot{\phi})\cdot h + D_{\phi}L(\phi,\dot{\phi})\cdot h$$

$$= \int_M \partial_{\dot{\phi}}\mathcal{L}(\phi,\dot{\phi},D\phi)\dot{h}\,d\mu + \left(\int_M \partial_{\phi}\mathcal{L}(\phi,\dot{\phi},D\phi)\cdot h\,d\mu + \int_M \partial_{D\phi}\mathcal{L}(\phi,\dot{\phi},D\phi)\cdot Dh\,d\mu\right).$$

Consider now Lagrange's equations derived above:

$$\frac{d}{dt} D_{\dot{\phi}}L(\phi,\dot{\phi}) = D_{\phi}L(\phi,\dot{\phi}).$$

This means that for any section h, the relation

$$\frac{d}{dt}\int \partial_{\dot{\phi}}\mathcal{L}(\phi,\dot{\phi},D\phi)\cdot h\,d\mu = \int \partial_{\phi}\mathcal{L}(\phi,\dot{\phi},D\phi)\cdot h\,d\mu + \int \partial_{D\phi}\mathcal{L}(\phi,\dot{\phi},D\phi)\cdot Dh\,d\mu$$

holds. This is exactly the weak form of the field equations. To get the strong form, assume h has compact support or vanishes on ∂M and that we have enough differentiability for the second integral on the right hand side to be integrated by parts. We then get

$$\int_M \left(\frac{\partial}{\partial t}\partial_{\dot\phi}\mathcal{L}\right)\cdot h\, d\mu = \int_M \{\partial_\phi \mathcal{L} - \operatorname{div}\partial_{D\phi}\mathcal{L}\}\cdot h\, d\mu .$$

Since h is arbitrary, we must have the Lagrangian density equation

$$\frac{\partial}{\partial t}(\partial_{\dot\phi}\mathcal{L}) = \partial_\phi \mathcal{L} - \operatorname{div}\partial_{D\phi}\mathcal{L}.$$

The expression on the right hand side is often called the functional derivative of $\mathcal{L}$ and is denoted

$$\frac{\delta\mathcal{L}}{\delta\phi} = \partial_\phi\mathcal{L} - \operatorname{div}\partial_{D\phi}\mathcal{L}.$$

The above formalism is suitable for field theories in all of space or with Dirichlet boundary conditions, i.e., in elasticity, the displacement problem. For the traction problem however, integration by parts leaves a boundary term so we shall have to modify things, as follows.

On using elasticity terminology for general field theory, let

$$T = \frac{\partial\mathcal{L}}{\partial D\phi}$$

be the first Piola-Kirchhoff stress tensor. Suppose, instead of ϕ being prescribed on ∂M, the traction $T\cdot N = \tau$ is prescribed. Here we assume the fibers E_x are embedded in a linear space so this makes sense, and N is the unit outward normal to ∂M. Now consider the Lagrangian

$$L(\phi,\dot\phi) = \int_M \mathcal{L}(\phi,\dot\phi,D\phi)\,d\mu + \int_{\partial M}\phi\cdot\tau\, da.$$

Then the same procedure as above shows that (if we have enough differentiability to pass through the weak form of the equations) Lagrange's equations

$$\frac{d}{dt} D_{\dot{\phi}} L(\phi,\dot{\phi}) = D_{\phi} L(\phi,\dot{\phi})$$

are equivalent to the field equations

$$\frac{\partial}{\partial t} \partial_{\dot{\phi}} \mathcal{L} = \partial_{\phi} \mathcal{L} - \operatorname{div} \partial_{DQ} \mathcal{L} \qquad \text{in} \quad M$$

<u>and</u> the boundary conditions

$$T \cdot N = \tau \qquad \text{on} \quad \partial M.$$

Thus we conclude that with this modified Lagrangian, the boundary conditions of traction emerge as part of Lagrange's equations. (This fact is important in numerical work since $T \cdot N = \tau$ are, in general, nonlinear functions of ϕ which are hard to impose in practice; cf. Oden and Reddy [205].)

We now consider Noether's theorem. As we saw above we can establish conservation laws for the total energy or other quantities under rather general circumstances. Noether's theorem gives results which are <u>formally</u> stronger (and really are stronger if we assume enough differentiability).

We begin by proving a conservation law for the energy density.

Let $\mathcal{L} : E \oplus J^1(E) \longrightarrow \mathbb{R}$ be a smooth Lagrangian density and let $\phi(t)$ be a differentiable curve of sections of E such that the Lagrange density equation of motion holds:

$$\frac{\partial}{\partial t}(\partial_{\dot{\phi}} \mathcal{L}) = \partial_{\phi} \mathcal{L} - \operatorname{div} \partial_{D\phi} \mathcal{L}.$$

Define the energy density by $\mathcal{E} = \dot{\phi} \partial_{\dot{\phi}} \mathcal{L} - \mathcal{L}$. Then $\mathcal{E}$ obeys the conservation equation ("continuity equation")

$$\frac{\partial \mathcal{E}}{\partial t} + \operatorname{div}(\dot{\phi}\partial_{D\phi}\mathcal{L}) = 0.$$

Indeed, using the chain rule together with the equation of motion, we find

$$\frac{\partial \mathcal{E}}{\partial t} = \frac{\partial}{\partial t}(\dot{\phi}\partial_{\dot{\phi}}\mathcal{L}) - \frac{\partial \mathcal{L}}{\partial t}$$

$$= \ddot{\phi}\partial_{\dot{\phi}}\mathcal{L} + \dot{\phi}\frac{\partial}{\partial t}(\partial_{\dot{\phi}}\mathcal{L}) - \ddot{\phi}\partial_{\dot{\phi}}\mathcal{L} - \dot{\phi}\partial_{\phi}\mathcal{L} - D\dot{\phi}\cdot\partial_{D\phi}\mathcal{L}$$

$$= \dot{\phi}\frac{\partial}{\partial t}(\partial_{\dot{\phi}}\mathcal{L}) - \dot{\phi}\partial_{\phi}\mathcal{L} - D\dot{\phi}\cdot\partial_{D\phi}\mathcal{L}$$

$$= \dot{\phi}\{\partial_{\phi}\mathcal{L} - \operatorname{div}\partial_{D\phi}\mathcal{L}\} - \dot{\phi}\partial_{\phi}\mathcal{L} - D\dot{\phi}\cdot\partial_{D\phi}\mathcal{L}$$

$$= -\dot{\phi}\operatorname{div}\partial_{D\phi}\mathcal{L} - D\dot{\phi}\cdot\partial_{D\phi}\mathcal{L}$$

$$= -\operatorname{div}(\dot{\phi}\partial_{D\phi}\mathcal{L}).$$

One can similarly localize the conservation laws associated with general symmetries. This proceeds, briefly, as follows. Let Ψ_t be a flow on M and let $\tilde{\Psi}_t$ be a flow of G on E, preserving fibers, covering Ψ_t. This extends, naturally to a flow on $J^1(E)$, called say $\tilde{\tilde{\Psi}}_t$. (It is determined by: $\tilde{\tilde{\Psi}}_t \circ j(\phi) \circ \Psi_t^{-1} = j(\tilde{\Psi}_t \circ \phi \circ \Psi_t^{-1})$ for ϕ a smooth section of E.) Let ξ_M and ξ_E be the corresponding infinitesimal generators on M and E. Assume that $\mathcal{L}$ is invariant in the sense that $\mathcal{L}\mu$ is unchanged under pull-back:

$$\mathcal{L} \circ (\tilde{\Psi}_t \oplus \tilde{\tilde{\Psi}}_t)\Psi_t^{*}\mu = \mathcal{L}\mu.$$

If ϕ is a solution of the Lagrange density equations, set

$$T_{\xi} = \mathcal{L}\xi_M + \partial_{D\phi}\mathcal{L}\cdot(\xi_E \circ \phi - D_{\phi}\cdot\xi_M) \quad \text{(a vector field on M)},$$

and

$$\mathfrak{J}_\xi = \partial_{\dot\phi}\mathcal{L}\cdot(\xi_t\circ\phi - D_\phi\cdot\xi_M) \quad \text{(a scalar field on } M\text{)},$$

where $\mathcal{L}$ stands for $\mathcal{L}(\dot\phi,\phi,D\phi)$ etc. Noether's theorem states that the following identity holds:

$$\frac{\partial\mathfrak{J}_\xi}{\partial t} + \operatorname{div} T_\xi = 0,$$

and it is readily proved along the lines indicated in the proof of the conservation law for the energy density.

10 THE HAMILTONIAN STRUCTURE OF NONLINEAR ELASTODYNAMICS

In this section we show that elastodynamics in the hyperelastic case is an infinite dimensional Hamiltonian system. For thermal or viscous effects, it becomes a Hamiltonian system with dissipation, as in the linear case.

Two applications of the Hamiltonian structure will be given. First of all, we show how the conservation laws of Knowles and Sternberg [158] and Fletcher [89] can be obtained using standard conservation theorems for Hamiltonian systems. Secondly, we show how conditions of incompressibility can be dealt with using ideas of Ebin and Marsden [77] and Ebin [76].

We begin with a description of the Hamiltonian structure. This consists of putting the standard variational methods in the geometric setting described in the previous section. (See Oden and Reddi [205] and Duvaut and Lions [73] for the traditional variational approach and Blancheton [27] and Chevallier [40] for work related to ours.)

As usual, we let M denote the reference configuration of the body. We assume M is a smooth compact manifold with boundary. For technical reasons related to smoothness, we shall deal with the displacement and traction problems separately and not with mixed boundary conditions.

Let $\mathscr{K}$ be the set of all regular configurations $\phi : M \longrightarrow N$ of class

$W^{s,p}$, $s > \frac{n}{p} + 1$ which, in the case of the displacement problem, have their assigned values on ∂M. As described in Section 4, $\mathscr{X}$ will be a smooth manifold. Its tangent space at ϕ, as shown in that section, consists of vector fields $u : M \longrightarrow TM$ covering ϕ; i.e., $u(X) \in T_{\phi(X)}N$. (Compare the definition of the bundle $\dot{E}$ in the previous section.)

Consider the equations of motion derived in Section 2:

$$\rho_0 A = \mathrm{DIV}\, T + \rho_0 B;$$

we assume $T = \rho_0(\partial W/\partial F)$ for a stored energy function W, where $F = T\phi$ is the displacement gradient, as usual. Now define the potential energy† $V : \mathscr{X} \longrightarrow \mathbb{R}$ by

$$V(\phi) = \int_M \rho_0 W(F) dV + \int_M \rho_0 B \cdot \phi dV$$

$$\left(+ \int_{\partial M} \tau \cdot \phi dA \quad \text{for the traction problem}\right).$$

(Here we explicitly assume $N = \mathbb{R}^n$ so that $\tau : \partial M_2 \longrightarrow \mathbb{R}^n$ and $\tau \cdot \phi$ and $B \cdot \phi$, the inner product, makes sense.)

Define the kinetic energy $K : T\mathscr{X} \longrightarrow \mathbb{R}$ by

$$K(u) = \frac{1}{2}\int_M \rho_0 \|u\|^2 dV,$$

and let $H = K + V$, $L = K - V$, and $\mathcal{L}(\phi,\dot{\phi},D\phi) = \frac{1}{2}\rho_0\|\dot{\phi}\|^2 - \rho_0 W(D\phi) - \rho_0\mathcal{V}_B$ as usual. The Hamiltonian system corresponding to this energy function is defined by Lagrange's equations:

---

†The term $B \cdot \phi$ appears a little strange but it is because of the dead loading. If b were prescribed, we would use $\mathcal{V} \circ \phi$ where $\mathcal{V}$ is a potential for b.

$$\frac{d}{dt}D_{\dot\phi}L(\phi,\dot\phi) = D_\phi L(\phi,\dot\phi).$$

As explained in the previous section, these equations are precisely the weak form of the equations of motion (and boundary conditions in the traction case).

We shall now use Noether's theorem to derive conservation laws for elasticity. From the Hamiltonian point of view, this is straightforward. (We have already considered the equation for energy in the previous section, so it can be omitted.)

In carrying this out, it is important to keep straight spatial and material invariances; i.e., invariances under transformations of N (the space) and of M (the body) respectively. Such ideas are implicit in work of Arnold [9] for instance, and it is in this respect that our treatment differs from that of Knowles, Sternberg [158] and Fletcher [89]. (These authors prove more in the cases they consider. They show that the only transformations which produce the desired infinitesimal invariance of $\mathcal{L}\mu$ are those with which they started.†)

Let us begin with spatial invariance. Let Ψ_t be a flow on N generated by a vector field w. This gives a flow on the bundle $M \times N \longrightarrow M$ by holding M pointwise fixed and moving N by Ψ_t. Thus, invariance of $\mathcal{L}$ reads:

$$\mathcal{L}(X,\Psi_t(\phi),D\Psi_t(\phi)\cdot\dot\phi,\ D\psi_t\cdot F)J(\psi_t) = \mathcal{L}(X,\phi,\dot\phi,F),$$

as an identity on $\mathcal{L}$ in its arguments X, ϕ, $\dot\phi$, F. Noether's theorem now

---

†These conservation laws can also be carried out for plates and shells in an analogous manner. See Naghdi [194] and Hughes and Marsden [132]. Notice that the results here include, as special cases, the linear case and the static case.

states that if ϕ satisfies the equations of motion, then we have the identity

$$\frac{\partial}{\partial t}(\partial_{\dot\phi}\mathcal{L}\cdot w) + \mathrm{DIV}(\partial_F\mathcal{L}\cdot w) = 0;$$

i.e.,

$$\frac{\partial}{\partial t}\left(\frac{\partial\mathcal{L}}{\partial\dot\phi^a}w^a\right) + \left(\frac{\partial\mathcal{L}}{\partial(\partial\phi^a/\partial X^A)}w^a\right)\Big|_A = 0.$$

Using the equations of motion, this can be simplified to

$$\partial_{\dot\phi}\mathcal{L}\cdot Dw\cdot\dot\phi + \partial_F\mathcal{L}\cdot Dw\cdot F = 0;$$

i.e.,

$$\frac{\partial\mathcal{L}}{\partial\dot\phi^a}w^a{}_{|b}\,\dot\phi^b + \frac{\partial\mathcal{L}}{\partial F^a{}_A}w^a{}_{|b}\,F^b{}_A = 0.$$

(In the notation of the previous section, $\xi_M = 0$, and $\xi_E = (0,w)$)

For $N = \mathbb{R}^3$ and choosing

(i) Ψ_t an arbitrary translational flow: $\Psi_t(x) = x + tw$, w a constant vector, we recover the equations of motion

$$\frac{\partial}{\partial t}\frac{\partial}{\partial\dot\phi} + \mathrm{DIV}\,\partial_F\mathcal{L} = 0,$$

i.e., balance of momentum. (The invariance of $\mathcal{L}$ does not depend on the point values of ϕ; cf. Section 3.)

(ii) Ψ_t an arbitrary flow of rotations; here $w(x) = Bx$ where B is an arbitrary skew symmetric matrix. Noether's theorem (together with the equations of motion) now gives:

$$\partial_{\dot\phi}\mathcal{L}\otimes\dot\phi + \partial_F\mathcal{L}\cdot F$$

is symmetric. For elasticity, $\partial_{\dot\phi}\mathcal{L}\otimes\dot\phi = \frac{\partial\mathcal{L}}{\partial\dot\phi^a}\dot\phi^b = \dot\phi^a\dot\phi^b$ is symmetric, so this reduces to the assertion that $t = TF$ is symmetric; i.e., balance of moment of momentum. Again, this invariance assumption will hold if $\mathcal{L}$ depends only on F through C, as in Section 3.

Notice that Noether's theorem provides a natural link between balance laws and material frame indifference. The assumption of material frame indifference plus the above Hamiltonian structure implies the usual balance laws. Thus, from an abstract point of view, the foundations of elasticity theory written in terms of a Lagrangian (or Hamiltonian) field theory seems somewhat more satisfactory - certainly more covariant - than the usual balance laws.

Next we turn to material invariance and let Λ_t be a flow on M generated by a vector field W on M. This induces a flow on the bundle $M \times N$ by merely holding N pointwise fixed. An important remark to be made is that Noether's theorem is purely local. Thus we may consider rotations about each point of M but restrict attention to a ball centered at each such point. The result of Noether's theorem is still valid since the proof is purely local. This is necessary since we wish to speak of isotropic materials without assuming M itself is invariant under rotations.

Invariance of $\mathcal{L}$ under Λ_t now reads as follows:

$$\mathcal{L}(\Lambda_t(X),\dot{\phi},\mathring{\phi},D\phi\cdot D\Lambda_t)J(\Lambda_t) = \mathcal{L}(X,\phi,\dot{\phi},D\phi)$$

as an identity on $\mathcal{L}$ in its arguments.

Noether's theorem in this case states that

$$\frac{\partial}{\partial t}(\partial_{\dot{\phi}}\mathcal{L}\cdot D\phi\cdot W) + \mathrm{DIV}(\partial_F\mathcal{L}\cdot D\phi\cdot W - \mathcal{L}W) = 0,$$

i.e.,

$$\frac{\partial}{\partial t}\left(\frac{\partial\mathcal{L}}{\partial\dot{\phi}^a}F^a{}_A W^A\right) + \left(\frac{\partial\mathcal{L}}{\partial F^a{}_A}F^a{}_B W^B - \mathcal{L}W^A\right)\Big|_A = 0.$$

(In the terminology of Noether's general theorem of the previous section, $\xi_M = W$, $\xi_E = (W,0)$. Note that the "field values" of ξ_E are zero.)

Again this can be rewritten using the equations of motion, if desired.

Now assume M is open in $\mathbb{R}^3$ and:

(i) Λ_t is an arbitrary translation $\Lambda_t(X) = X + tW$, W a constant vector. Then $\mathcal{L}$ will be invariant if it is <u>homogeneous</u>, i.e., independent of X. In this case, Noether's theorem yields the identity:

$$\frac{\partial}{\partial t}\left(\frac{\partial \mathcal{L}}{\partial \dot{\phi}^a} F^a{}_A\right) + \left(\frac{\partial \mathcal{L}}{\partial F^a{}_B} F^a{}_A\right)_{|B} - \mathcal{L}_{|A} = 0,$$

i.e., for any subbody $U \subset M$, with unit outward normal N_A,

$$\frac{\partial}{\partial t}\int_U \frac{\partial \mathcal{L}}{\partial \dot{\phi}^a} F^a{}_A dV = \int_{\partial U} (\mathcal{L}N_A - \frac{\partial \mathcal{L}}{\partial F^a{}_B} F^a{}_A N_B) dA.$$

This identity expresses conservation of momentum; indeed, for elasticity,

$$\frac{\partial \mathcal{L}}{\partial \dot{\phi}^a} F^a{}_A = \rho_0 \dot{\phi}_a F^a{}_A$$

is just the momentum density expressed in material coordinates. Thus,

$$\mathcal{L}N_A - \frac{\partial \mathcal{L}}{\partial F^a{}_B} F^a{}_A N_B = \mathcal{L}N_A - T^B{}_a F^a{}_A N_B,$$

(where T is the first Piola-Kirchhoff stress tensor) may be interpreted as a momentum flux. (If $\mathcal{L}$ is independent of X, the momentum identity can be verified directly using the equations of motion and the chain rule on $\mathcal{L}_{|A}$.)

(ii) Λ_t is a rotation about the point X_0, then $W = B(X - X_0)$ where B is an arbitrary skew symmetric matrix. We can write, in vector notation,

$$W = V \times (X - X_0),$$

where V is a constant vector, or

$$W_A = \varepsilon_{ABC} V^B (X^C - X_0^C),$$

where ε^{ABC} is the alternator. Noether's theorem becomes (in Euclidean coordinates):

$$\frac{\partial}{\partial t}\left(\frac{\partial \mathcal{L}}{\partial \dot{\phi}^a} F^a_A \varepsilon^{ABC} X_C\right) + \left(\frac{\partial \mathcal{L}}{\partial F^a_D} F^a_A \varepsilon^{ABC} X_C - \mathcal{L}\varepsilon^{DBC} X_C\right)\Big|_D = 0.$$

Here this expresses a conservation law for the actual angular momentum of the body. For it to hold $\mathcal{L}$ must be isotropic in the sense discussed in Section 3.

If $\mathcal{L}$ is also homogeneous, then using the identity in (i), (ii) reduces to

$$\frac{\partial \mathcal{L}}{\partial F^a_D} F^a_A \varepsilon^{ABD} = 0,$$

i.e.,

$$T_{aD} F^a_A \varepsilon^{ABD} = 0, \qquad B = 1,2,3.$$

If one uses the standard isotropic representation for T, one sees directly that this identity holds.

Remarks 1. All of this can equally well be done from a space-time point of view.

2. Other symmetry groups (e.g., dilatations, etc.,) can be dealt with in the same way.

As our second application of Hamiltonian methods, we shall discuss incompressible elasticity. Here we impose the constraint $J = 1$; i.e., $\operatorname{div} v = 0$, and replace the Cauchy stress t by $t + pI$ where $JtF^{-1} = \rho_0 \frac{\partial W}{\partial F}$, that part of the stress derived from a stored energy function and where p is determined by the incompressibility condition. Interestingly, the geometric ideas developed in the last section can be of technical benefit for the incompressible case. One of our goals now is to explain why this is so. The ideas for this program are due to Ebin and Marsden [77] for the case of fluids.

As we have seen, the equations of elastodynamics may be regarded as a Hamiltonian system with configuration space $\mathscr{C}$. For incompressible elasticity we work with

$$\mathscr{C}_\mu = \{\phi \in \mathscr{C} \mid J(\phi) = 1\}.$$

Recall that in the displacement problem ϕ is fixed on ∂M, but no boundary conditions are imposed on ϕ for the traction problem.

To distinguish the various spaces, we introduce the notation

$$\mathscr{C}_\mu = \{\phi : M \longrightarrow N \mid \phi \text{ is regular and } J(\phi) = 1\},$$

$$\mathscr{C}_{\mu,\parallel} = \{\phi : M \longrightarrow N \mid \phi \text{ is regular, } J(\phi) = 1 \text{ and } \phi(\partial M) \subset \partial M\},$$

$$\mathscr{C}_{\mu,0} = \{\phi : M \longrightarrow N \mid \phi \text{ is regular, } J(\phi) = 1 \text{ and } \phi(X) = X \text{ for } X \in \partial M\}.$$

<u>Proposition</u> In $W^{s,p}$, $s > \frac{n}{p} + 1$, each of these spaces is a smooth submanifold:

$$\mathscr{C}_{\mu,0} \subset \mathscr{C}_{\mu,\parallel} \subset \mathscr{C}_\mu \subset \mathscr{C}.$$

Their tangent spaces are:

$$T_\phi \mathscr{C}_\mu = \{u \in T_\phi \mathscr{C} \mid \operatorname{div}(u \circ \phi^{-1}) = 0\},$$

$$T_\phi \mathcal{H}_{\mu,\|} = \{u \in T_\phi \mathcal{H} \mid \operatorname{div}(u \circ \phi^{-1}) = 0 \text{ and } u \text{ is parallel to } \partial M\},$$

$$T_\phi \mathcal{H}_{\mu,0} = \{u \in T_\phi \mathcal{H} \mid \operatorname{div}(u \circ \phi^{-1}) = 0 \text{ and } u = 0 \text{ on } \partial M\}.$$

The proof relies on the Hodge decomposition. Let $\mathbb{P}_\|$ denote the projection of a vector field onto its divergence free part parallel to the boundary and $\mathbb{P}$ be the projection onto the divergence free part, corresponding to the orthogonal decompositions:

$$u = v + \nabla p, \quad \operatorname{div} v = 0, \quad v \| \partial M,$$

and

$$u = v + \nabla p, \quad \operatorname{div} v = 0, \quad p = \text{constant on } \partial M,$$

respectively.

The proof that $\mathcal{H}_\mu$ and $\mathcal{H}_{\mu,\|}$ are submanifolds using these decompositions is given in Ebin-Marsden [77] and need not be repeated. (For $\mathcal{H}_{\mu,0}$ one considers the map

$$B : \mathcal{H}_{\mu,\|} \longrightarrow W^{s-(1/p),p}(\partial M, \partial M),$$

$$\phi \mapsto \phi \mid \partial M,$$

and shows it is a submersion).

The appropriate phase space for incompressible elasticity is $\mathcal{H}_\mu$, or $\mathcal{H}_{\mu,\|}$ for the displacement problem. The space $\mathcal{H}_{\mu,0}$ will appear as the domain of the generator in the displacement problem along with other compatibility conditions depending on the degree of smoothness assumed.

Proposition The equations (and boundary conditions) of incompressible elasticity are equivalent to Lagrange's equations for the usual Lagrangian

$$L(\phi,\dot{\phi}) = \frac{1}{2}\int_M \rho_0 \|\dot{\phi}\|^2 dV - \int_M \rho_0 W(F)\, dV - \int_M \rho_0 \mathcal{V}_b \circ \phi \, dV$$

$$\left(- \int_{\partial M} \tau \cdot \phi \, dV \quad \text{for the traction problem}\right)$$

defined on $T\mathscr{H}_{\mu,\|}$ (or $T\mathscr{H}_{\mu}$ for the traction problem).

The proof of this proceeds in the usual way outlined earlier.

Now define the bundle map

$$P : T\mathscr{H} \mid \mathscr{H}_{\mu} \longrightarrow T\mathscr{H}_{\mu},$$

$$u \longmapsto \mathbb{P}(u \circ \phi^{-1}) \circ \phi,$$

(and similarly on $\mathscr{H}_{\mu,\|}$).

Theorem P is a smooth mapping from the $W^{s,p}$ topology to itself, $s > \frac{n}{p} + 1$.

This result is at the heart of the paper of Ebin and Marsden [77]. Indeed, it shows that the spray on the space $\mathscr{H}_{\mu,\|}$ is smooth and so the local existence and uniqueness of geodesics becomes trivial. This is one way of proving the existence and uniqueness theory for the Euler equations for an incompressible fluid.

For elasticity, we know that the equations can be written, using the results of Section 9, as

$$\frac{\partial^2 \phi}{\partial t^2} = TP \circ S(\phi,\dot{\phi}) - [P \operatorname{grad} V(\phi)]^{\ell}$$

where S is the spray - the algebraic Christoffel symbol terms in the equations of motion. Since P is smooth we see that these equations differ from the compressible equations

$$\frac{\partial^2\phi}{\partial t^2} = S(\phi,\dot{\phi}) - [\operatorname{grad} V(\phi)]^{\ell}$$

by the application of the smooth map P.

For the case of elasticity with boundary conditions the initial value problem is not yet completely settled and is very complex. Kato [142] has proved local existence and uniqueness, but not continuous dependence on initial data. (We shall remark on this in the next section.) For all of space however, Hughes, Kato and Marsden [129] have shown local existence, uniqueness and continuous dependence on the initial data in $H^{s+1} \times H^s$ spaces, $s > \frac{n}{2}$ (again see the next section).

If those proofs are examined, it will be seen from the fact that P is smooth (so P and its derivative are Lipschitz from H^s to H^s), that the estimates needed in the abstract existence theorem are not damaged.

Therefore one can conclude that for the equations of incompressible elasticity in all of space, under the usual strong ellipticity condition we have local existence, uniqueness and continuous dependence on the initial data in $H^{s+1} \times H^s$.

We conclude with a few remarks on the incompressible limit, based on Rubin and Ungar [223] and Ebin [76]. If we imagine the incompressible pressure p replaced by a constitutive law $p_k(\rho)$, where $(dp_k/d\rho) = k$, so $1/k$ is the compressibility, then a potential V_k is added to our Hamiltonian which has the property:

$$V_k(\phi) = 0 \quad \text{if} \quad \phi \in \mathscr{K}_\mu,$$

$$V_k(\phi) \underset{k\to\infty}{\longrightarrow} \infty \quad \text{if} \quad \phi \notin \mathscr{K}_\mu.$$

In such a case, it is intuitively clear from conservation of energy that this ought to force compressible solutions with initial data in $\mathscr{K}_\mu$ to converge to the incompressible solutions as $k \longrightarrow \infty$.

Such convergence in the linear case can be proved by the Trotter-Kato theorem (see Section 5). Kato [138] has proved analogues of this for nonlinear equations which, following Hughes, Kato and Marsden [129], are applicable to nonlinear elastodynamics. These approximation theorems may now be used in the proofs given by Ebin [76]. Although all the details have not been checked, it seems fairly clear that this enables one to prove the convergence of solutions in the incompressible limit, at least for short time.

We also mention that the smoothness of the projection P and the convergence of the constraining forces as $k \longrightarrow \infty$ should enable one to give a simple proof of convergence of solutions of the stationary solutions by merely employing the implicit function theorem. (Rostamian [222] gives some complementary results in this direction but in larger function spaces than may be appropriate for the methods just outlined.)

11 A SURVEY OF SELECTED NONLINEAR PROBLEMS

In this final section we shall describe briefly a few problems in nonlinear elasticity which we feel are fundamental. Needless to say, there is a vast number of basic issues not described here. Qualitative and global problems in bifurcation theory and global existence of weak solutions (even in one dimension) are just two of these.

We shall consider first some aspects of nonlinear elastostatics and then move on to elastodynamics.

1. (Minimizers and Global Analysis) The existence of minimizers for the total energy in hyperelastostatics has been demonstrated by Ball [12,15] under conditions which do not imply uniqueness of solution. The usual

convexity arguments (cf. Beju [24]) are unsatisfactory because they preclude non-uniqueness and hence buckling. However many basic problems remain:

(a) Are the minimizers found by Ball regular enough to be called solutions?

(b) Can one develop a Morse theory which enables one to count solutions algebraically?

The usual Morse theory, as developed by Palais and Smale (see, e.g., Schwartz [225]) is suitable for semilinear problems, but is not applicable for many quasi-linear problems such as elasticity (The same can be said for existing topological methods in bifurcation theory, such as Rabinowitz [215]; however as Antman and Rosenfeld [8] have shown, some quasilinear problems can be made to be semilinear if a clever change of variables is made). The work of Tromba [242] on the Plateau problem seems to be an important step towards dealing with these questions. For semi-linear approximations a great deal is known. The von Karmen equations for a plate is a good example. See, for instance Berger [25] and Hale [118].

2. (Embedding in the Dynamics) It should be borne in mind that in elastostatics one is studying the fixed points of a dynamical system and that the problem may not be considered completely solved until the dynamical structure enveloping these points is understood, i.e., are the points dynamically stable, unstable or saddles? Are there periodic orbits nearby, etc. (Remark 5(a) below is relevant to this discussion.) Again for semilinear approximations, a great deal is known. See, for instance, Reiss and Matkowsky [218], Ball [11] and Holmes and Marsden [128].

3. (Stoppeli's Results On the Traction Problem With an Axis of Equilibrium.) A method for analyzing this problem will only briefly be sketched here, with details given elsewhere (Chillingworth and Marsden [42]). In Section 8 it

was explained that in the traction problem where the loads have no axis of equilibrium and are sufficiently small, the nonlinear problem has, locally, a unique solution. For the case of an axis of equilibrium, Stoppelli [236] has shown that there may to 0, 1, 2 or 3 solutions (under suitable hypotheses) and he showed how to expand the solutions in power series in θ, $\sqrt{|\theta|}$ or $\sqrt[3]{|\theta|}$. Actually, if the hypotheses are relaxed further, it seems that using the methods below, one can get up to nine solutions near equilibrium (e.g., four saddles, a maximum and four minima).

Our point of view is to use the methods of Catastrophe theory and generic bifurcation theory. Let $\mathscr{K}$ be the space of configurations of a body M and eliminate translations by taking, say $\phi(0) = 0$. Define, following the Hamiltonian ideas in Section 10,

$$V : \mathscr{K} \longrightarrow \mathbb{R}$$

$$\phi \longmapsto \int_M \rho_0 W(F) dV + \int_M \rho_0 B \cdot \phi dV + \int_{\partial M} \tau \cdot \phi dA.$$

Then the solutions of the traction problem are exactly the critical points of V. Catastrophe theory is precisely suited for such a situation, and its application to elasticity is not unusual (see Chillingworth [41] and Zeeman [263]).

We can regard W, B and τ as parameters in V and can seek to find the structure of the set of critical points of V. Let us write, therefore,

$$V(\phi;\ W, B, \tau).$$

Material frame indifference tells us that if Q is orthogonal,

$$V(Q\phi, W, B, \tau) = V(\phi, W, Q^{-1}B, Q^{-1}\tau),$$

so V has an equivariance property with respect to the action of the orthogonal group on $\mathscr{H}$ and the parameters, which must be taken into account.

The condition that B and τ have no axis of equilibrium implies that in the appropriate quotient space, dV will have a unique critical point near equilibrium if B and τ are small. This procedure gives what seems to be a more geometrically transparent approach to Stoppelli's results. Of course, secretly the method is the same as that outlined in Section 8.

Next, suppose B, τ have an axis e of equilibrium. This time V has a degenerate critical point in the quotient space. By passing to the bifurcation equation we eliminate the "trivial" directions and obtain a problem in one variable (corresponding to assuming that the axis e is "non-degenerate").

We make the generic assumption that the Taylor expansion of W as a function of the Cauchy-Green tensor C has the form

$$W(C) = aC + \gamma C^2 + \ldots$$

where $\gamma \neq 0$.

Now we hold everything fixed except a and θ (a scaling parameter in $B = \theta B_0$, $\tau = \theta\tau_0$). Then V has the form of a quartic:

$$V(x) = \gamma x^4 + ax^2 + \theta x + \text{h.o.t.}$$

Its critical points form a cusp, as shown in Fig. 12.

For various values of a and varying θ note that we get either one or three solutions. (If the axis e is rotated, the picture will rotate accordingly.) For example, if we pass through the origin along the θ axis, we get a unique solution whose amplitude varies with $\sqrt[3]{|\theta|}$.

Using these methods we can reproduce Stoppelli's results. Moreover, it is expected that if there are two axes of equilibrium or a degenerate axis of

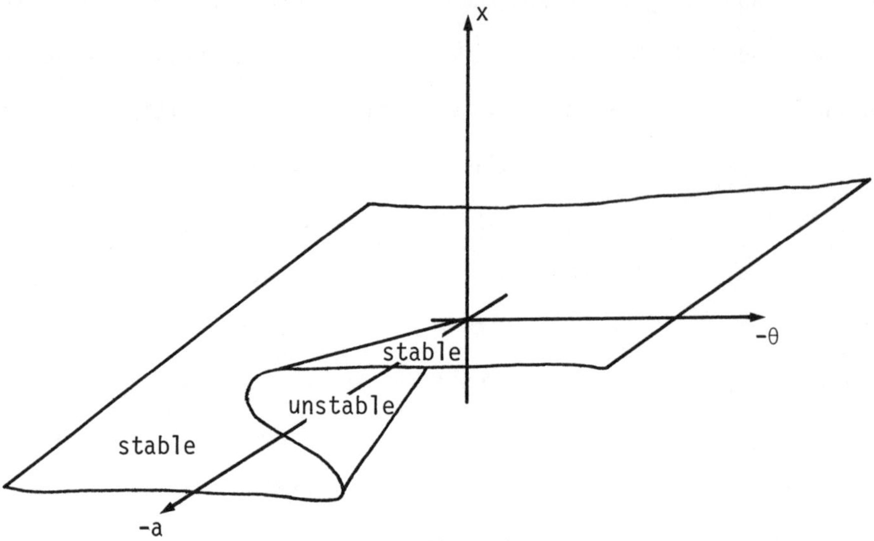

Fig. 12.

equilibrium one obtains a double cusp in which up to nine solutions are possible. This is intended to complete Stoppelli's classification of the structure of the set of solutions.

In the language of §4, the nonlinear equations in the presence of an axis of equilibrium are linearization unstable. As such, we know that any possible directions of perturbation must satisfy certain compatability conditions. This reproduces the Signorini compatibility conditions in a perturbation expansion (See Truesdell and Noll [248], p.226†).

If one views the equilibrium solutions in the context of the dynamics, one sees that the linearization instability in elastostatics disappears; i.e., there is no reason to reject either linear or nonlinear elastostatics; the fact is that the linearized theory does not give solutions which to first

---

†"In incompatible cases, the solution given by the infinitesimal theory cannot be the first term in a series... satisfying formally the particular problem of the finite theory". Compare with our discussion in Section 4.

order approximate solutions in the nonlinear theory. These points are discussed in Truesdell and Noll [248] and in Capriz and Podio Guidugli [33]. (The example of a nonlinearly bent plate discussed in the latter is useful for seeing what happens physically.)

4. (Well-Posedness of the equations of nonlinear elastodynamics) In Hughes, Kato and Marsden [129] the equations of nonlinear elastodynamics in all of $\mathbb{R}^n$ are shown to be locally well-posed in $X = H^{s+1} \times H^s$ for $s > \frac{n}{2} + 1$. This means that for short time, there exists unique solutions depending continuously on the initial data.

The idea is to consider an abstract quasi-linear Cauchy problem

$$\dot{u} + A(u)u = f(u),$$
$$u(0) = \phi.$$

One sets up a map of functions $u(t)$ with the given initial data to the solution of the frozen coefficient equations

$$\dot{v} + A(u(t))v = f(u(t)),$$
$$v(0) = \phi.$$

Note that we now deal with time dependent linear equations and for them the results of Kato [138] are used. If $A(u)$ is a generator for each u (which as verified in Section 7 is true for elasticity) and certain other technical conditions hold, one can show that this map has a unique fixed point which is the desired solution.

A delicate part of the analysis is the continuous dependence on the initial data. Indeed, as remarked in Section 9, this dependence is not in general locally Lipschitz.

One of the key features of the analysis is the use of several Banach spaces, $Y \subset X \subset Z$. Here for each $u \in Y$, $A(u)$ is a generator in Z of a quasi-contractive semigroup which leaves Y invariant (i.e., is Y-stable) and $D(A) \subset X$. One assumes there is an isomorphism

$$S(u) \in B(Y,Z)$$

such that

$$S(u)A(u)S(u)^{-1} = A(u) + B(u),$$

where $B(u) \in B(Z)$. This isomorphism S plays a fundamental role. The rest of the assumptions needed are more of a technical nature and we refer to the papers cited for details.

Smooth solutions will not exist globally in time because of shocks. Finding spaces to deal with this is a major open problem.

For the initial boundary value problem, Kato [142] has shown local existence and uniqueness, but the continuous dependence on the initial data is not yet known. The major complication which needs to be dealt with is the compatability conditions and the possibility that the domains of powers of A will be time dependent, even though that of A is not. All of this is caused by the degree of smoothness required in the methods.

We shall now sketch out an idea which may be useful in proving the continuous dependence. Write the equations this way:

$$\dot{u} + \mathcal{A}(u) = 0,$$
$$u_0 = \phi,$$

where $\mathcal{A}(u) = A(u)\cdot u$ and we have dropped the term $f(u)$ for simplicity. Suppose that the boundary conditions are written

$$B(u) = 0, \qquad u \in D(\mathcal{A}).$$

If we seek solutions in the domain of $[A(u)]^3$, then the compatability conditions for the initial data are obtained by differentiating $B(u) = 0$ twice:

(i) $B(\phi) = 0$,

(ii) $B'(\phi)\cdot\mathcal{A}(\phi) = 0$,

(iii) $B''(\phi)\cdot(\mathcal{A}(\phi),\mathcal{A}(\phi)) + B'(\phi)\cdot\mathcal{A}'(\phi)\cdot\mathcal{A}(\phi) = 0$.

The difficulty is that even if $B(\phi) = 0$ are linear boundary conditions, this is a nonlinear space of functions in which we seek the solution. Let C denote the space of functions satisfying the compatability conditions. It seems natural to try to show C is a smooth manifold. We assume B itself is <u>linear</u> for simplicity.

<u>Proposition</u> Let $\phi \in C$ and assume

(a) B is surjective,

(b) the linear boundary value problem

$$\mathcal{A}'(\phi)\cdot\psi = \rho; \quad B\cdot\psi = 0,$$

has a solution ψ for any ρ,

(c) the linear boundary value problem

$$\mathcal{A}''(\phi)(\mathcal{A}(\phi),\psi) + (\mathcal{A}'(\phi))^2\psi = \rho; \quad B\cdot\psi = 0, \quad B\cdot\mathcal{A}'(\phi)\cdot\psi = 0$$

has a solution ψ for any ρ [Note that in case $\mathcal{A}(\phi) = A(\phi)\cdot\phi$, this has highest order term $(A(\phi)^2\cdot\psi]$. Then C is a smooth manifold in the neighbourhood of ϕ.

<u>Proof</u>. Let $C_1 = \ker B$, the "first" boundary conditions. Map

$$\phi \in C_1 \longmapsto B'(\phi)\cdot\mathcal{A}(\phi) = B\cdot\mathcal{A}(\phi).$$

This has derivative at ϕ given by

$$\psi \longmapsto B\cdot\mathcal{A}'(\phi)\cdot\psi.$$

Since B is surjective and $\mathcal{A}'(\phi) : \ker B \longrightarrow$ (range space) is surjective by (b), this map has a surjective derivative, so

$$C_2 = \{\phi \in C_1 \mid B\cdot\mathcal{A}(\phi) = 0\}$$

is a submanifold of C_1 with tangent space $T_\phi C_2 = \{\psi : B\psi = 0$ and $B\cdot\mathcal{A}'(\phi)\cdot\psi = 0\}$, by the implicit function theorem.

Finally, map

$$C_2 \longrightarrow \text{(range space)}$$
$$\phi \longmapsto B\cdot\mathcal{A}'(\phi)\cdot\mathcal{A}(\phi)$$

which has derivative

$$\psi \longmapsto B\cdot(\mathcal{A}''(\phi)(\mathcal{A}(\phi),\psi) + (\mathcal{A}'(\phi))^2\cdot\psi).$$

Thus, by assumption (c), this is surjective on $T_\phi C_2$. Thus,

$$C_3 = \{\phi \in C_2 \mid B\cdot\mathcal{A}'(\phi)\cdot\mathcal{A}(\phi) = 0\} = C$$

is a submanifold by the implicit function theorem. □

We want to solve

$$\frac{du}{dt} + \mathcal{A}(u) = 0$$

for $u(t) \in C$ given initial condition $\phi \in C$. To do so, we can try using the local diffeomorphism

$$\Phi : C \longrightarrow (\text{linear space}) = Y$$

mapping a neighbourhood of ϕ in C to a ball in a linear space obtained from the proof above. (So Φ is only defined implicitly.) We write $\Phi(\phi) = \phi$.

Let $v = \Phi(u)$. Thus

$$\frac{dv}{dt} = \Phi'(u)\frac{du}{dt}$$

$$= \Phi'(\Phi^{-1}(v))\cdot(-\mathcal{A}(\Phi^{-1}(v))),$$

so v satisfies

$$\frac{dv}{dt} + \tilde{\mathcal{A}}(v) = 0$$

where $\tilde{\mathcal{A}}(v) = \Phi'(\Phi^{-1}(v))\cdot\mathcal{A}(\Phi^{-1}(v))$.

(In geometry notation,

$$\tilde{\mathcal{A}} = \Phi_*\mathcal{A}).$$

If the modified problem is well-posed, then clearly the original one is as well. We can choose $Y = \{\psi \mid B\psi = 0,\ B\cdot\mathcal{A}'(\phi)\cdot\psi = 0$ and $B\cdot\mathcal{A}''(\phi)(\mathcal{A}(\phi),\psi) + (\mathcal{A}'(\phi))^2\psi = 0\}$ and let Φ be the projection of C onto Y; see Fig. 13.

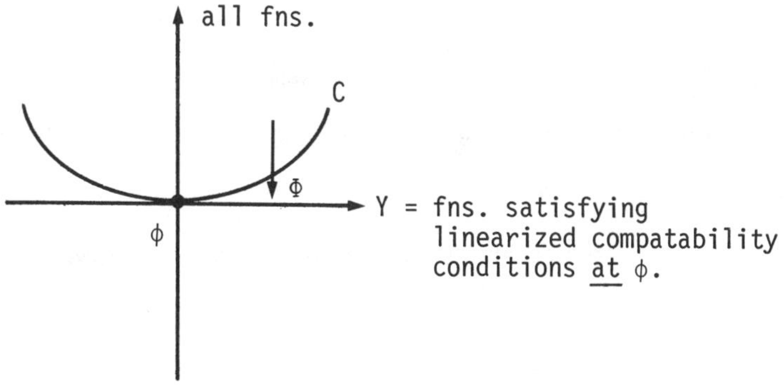

Fig. 13.

In the new formulation, v lies in the fixed linear space Y and so we can reasonably suppose that a fixed "S-operator" can be chosen.

Now let $\mathcal{A}(u) = A(u)\cdot u$. Since Φ is a linear projection,

$$\tilde{\mathcal{A}}(v) = \Phi\cdot A(\phi^{-1}(v))\cdot\Phi^{-1}(v)$$

$(\Phi^{-1}(v) \in C)$ is still quasi-linear. Since Φ projects onto Y,

$$\tilde{\mathcal{A}}(\phi) = A(\phi)\cdot\phi = \mathcal{A}(\phi)$$

so we should be able to take S to be the S that works for $A(\phi)$.

In this context it seems reasonable to ask that the abstract quasi-linear theorem in Hughes, Kato and Marsden [129] applies to the new system $\tilde{\mathcal{A}}$. If it did, one would be able to prove continuous dependence for the initial boundary value problem.

5. (<u>The Energy Criterion</u>) The energy criterion for elasticity, or more generally for a Hamiltonian system of the form $H(x,\dot{x}) = \frac{1}{2}\|\dot{x}\|^2 + V(x)$, states that if x_0 is a minimum for the potential energy, $(x_0,0)$ is an equilibrium point, then $(x_0,0)$ is dynamically stable.

This criterion has been discussed extensively in the literature. See, for instance, Koiter [164,165], Knops and Wilkes [157], Gurtin [110] and references therein. In particular, examples like $V(x) = \frac{1}{2}(u_X^2 - u_X^4)$ in one dimension (see Knops [146] and 5(a) below) suggest that for elasticity, such a sweeping criterion is probably false.

To obtain a rigorous result, two assumptions are usually made.

(S1) There is a continuous local semiflow (perhaps consisting of weak solutions) defined in a neighbourhood of $(x_0,0)$ in some Banach space such that (a) $H(x,\dot{x})$ decreases along solutions and

(b) there is an $\eta > 0$ and $\tau > 0$ such that initial data in the ball of radius η about $(x_0,0)$ can be continued forward in time by an amount at least τ;

(S2) x_0 lies in a potential well; i.e., there is an $\varepsilon > 0$ such that (a) $V(x_0) < V(x)$ if $0 < \|x - x_0\| < \varepsilon$, and

(b) $\rho(\varepsilon') =: \inf_{\|x-x_0\|=\varepsilon'} V(x) > V(x_0)$, for all $0 < \varepsilon' \leqslant \varepsilon$.

The following is then well known from the above references.

Theorem If (S1) and (S2) hold, then $(x_0,0)$ is dynamically stable (in the Liapunov sense).

Proof. We must show that if $\|\dot{x}(0)\|$ and $\|x(0) - x_0\|$ (these are different norms in general, but it is irrelevant here) are sufficiently small, then the solution can be continued for all time and remains in a neighbourhood of $(x_0,0)$. By the fact that energy is decreasing, we have

$$\tfrac{1}{2}\|\dot{x}(t)\|^2 + V(x(t)) \leqslant \tfrac{1}{2}\|\dot{x}(0)\|^2 + V(x(0)).$$

Choose $\frac{1}{2}\|\dot{x}(0)\|^2 < \min(\rho(\varepsilon') - V(x_0), \frac{1}{2}\delta^2)$ where $0 < 2\varepsilon' \leqslant \varepsilon$ and $\{(x,\dot{x}) \mid \|x - x_0\| < 2\varepsilon', \|\dot{x}\| < \delta\}$ is a ball about $(x_0,0)$ in which (S1)(b) holds. Let $[0,T)$ be the maximum interval for which the solution remains in the ball $\{(x,\dot{x}) \mid \|x - x_0\| < \varepsilon', \|\dot{x}\| < \delta\}$. Assume $T < \infty$. Then since $x(t)$, $\dot{x}(t)$ is defined for some $[0,T+\tau)$ and satisfies $\|x(t) - x_0\| < \varepsilon$, we have $V(x(t)) \geqslant V(x_0)$ and so $\|\dot{x}(t)\| \leqslant \|\dot{x}(0)\| < \delta$. Now this implies that $\|x(T) - x_0\| = \varepsilon'$. Hence,

$$\begin{aligned}\rho(\varepsilon') &\leqslant V(x(T)) \\ &\leqslant \tfrac{1}{2}\|\dot{x}(0)\|^2 + V(x(0)) \\ &< (\rho(\varepsilon') - V(x(0))) + V(x(0)) \\ &= \rho(\varepsilon').\end{aligned}$$

which is clearly a contradiction. Thus $T = +\infty$. □

Remark If we modify (S2)(b) by only assuming $\rho(\varepsilon') > V(x_0)$ for some $\varepsilon' > 0$, but assume in (S1) that η can be arbitrarily large (with τ

depending on η), and in particular $\eta > 2\varepsilon'$, then the same result is true.

5.(a) (An Example) As in Remark 4 above, there are lots of spaces in which one can get a local dynamical system, but virtually nothing is known about the validity of condition (S1)(b), even in the simplest examples. However one can run into serious trouble trying to verify (S2) as well, as Knops [146] and Knops and Payne [154] have shown. We shall illustrate the difficulty with a general example (Ball, Knops and Marsden [16]). We give the result for the interval [0,1] with zero boundary conditions for the displacement $u(x)$, but the argument is rather general. Let $u_0(x) = 0$ denote the trivial equilibrium solution for a potential

$$V(u) = \int_0^1 W(u_x)dx.$$

The result shows:

(a) positivity of the second variation (in $W^{1,2} = H^1$) at the trivial solution implies u_0 locally minimizes V in a topology as strong as $W^{1,\infty}$ although it need not locally minimize V in $W^{1,p}$ for any p, $1 \leqslant p < \infty$. In any topology as strong as $W^{1,\infty}$ we always have

$$\inf_{\|u-u_0\| = \varepsilon} V(u) = V(u_0)$$

for ε sufficiently small; i.e., (S2)(b) necessarily fails in topologies as strong as $W^{1,\infty}$.

Thus, the only spaces left are ones like $W^{1,p}$ where (S2)(b) is possible. Then (S2)(a) may be verifiable using suitable convexity assumptions (like convexity in one dimension and Ball's polyconvexity in higher dimensions). Of course this still leaves (S1) wide open.

For the details, we let $W : \mathbb{R} \longrightarrow \mathbb{R}$ be a smooth function with $W'(0) = 0$

and $W''(0) > 0$. As in the example in Section 8, u_0 is an extremal and the second variation of V at u_0 is positive definite. Let X be a Banach space continuously included in $W^{1,\infty}$. Then there is an $\varepsilon > 0$ such that

$$\text{if } 0 < \|u - u_0\|_X < \varepsilon \text{ then } V(u) > V(u_0).$$

i.e., u_0 is a strict local minimum for V. This follows trivially from the fact that 0 is a local minimum of W and that the topology on X is as strong as that of $W^{1,\infty}$.

In $W^{1,p}$ one cannot conclude that u_0 is a local minimum. Indeed the example $W(u_x) = \frac{1}{2}(u_x^2 - u_x^4)$ shows that in any $W^{1,p}$ neighbourhood, $V(u)$ can be unbounded below, even though its second variation at u_0 is positive definite.

Finally, we show that

$$\inf_{\|u-u_0\|_X = \varepsilon} V(u) = V(u_0).$$

Indeed, by Taylor's theorem,

$$\begin{aligned} V(u) - V(u_0) &= \int_0^1 (W(u_x) - W(0))dx \\ &= \int_0^1 \int_0^1 (1-s)W''(su_x)(u_x)^2 ds dx \\ &\leq C\int_0^1 (u_x)^2 dx, \end{aligned}$$

where $C > 0$, since su_x is uniformly bounded (by the assumption $X \subset W^{1,\infty}$) and W'' is continuous. However, the topology on X is strictly stronger than the $W^{1,2}$ topology, and so

$$\inf_{\|u-u_0\|_X = \varepsilon} \int_0^1 (u_x)^2 dx = 0.$$

This proves our claim.

5.(b) (Semilinear Approximations) In practice the energy criterion has great success, according to Koiter [165]. However, this is consistent with the possibility that the energy criterion may fail for elastodynamics. Indeed "in practice" one usually does not observe the very high frequency motions. Masking, dissipating or averaging them may amount to replacing the quasilinear equations of elastodynamics by finite dimensional or semilinear approximations†. For the latter, the validity of the energy criterion is basically trivial (see, for instance, Marsden [182] and Payne and Sattinger [209]). We shall illustrate the idea with the equations††

$$u_{tt} = (T(u_x))_x + u_{xxxx}$$

where $T(u_x) = W'(u_x)$, $W(0) = W'(0) = 0$, $W''(0) \geqslant 0$ and W is smooth. We work in one dimension on an interval with boundary conditions $u = 0$, $u_x = 0$ at the ends.

We claim that in $H^2 \times L_2 = X$ and $u_0 = 0$, the conditions (S1) and (S2) are true. The equations are clearly Hamiltonian with

$$H(u,\dot{u}) = \tfrac{1}{2}\int |\dot{u}(x)|^2 dx + \int W(u_x) dx + \tfrac{1}{2}\int |u_{xx}|^2 dx,$$

and

$$V(u) = \tfrac{1}{2}\int |u_{xx}|^2 dx + \int W(u_x) dx.$$

---

†The use of Fourier Integral Operators and Oscillatory Integrals may be relevant here.

††For a more complex example, see Holmes and Marsden [128]. This simple illustration was suggested by R. Knops.

First of all, the linear part; i.e., the operator

$$A' = \begin{pmatrix} 0 & I \\ A & 0 \end{pmatrix}, \quad A(u) = u_{xxxx},$$

generates a contraction semi-group on X, where X is given the linearized energy norm

$$\|u,\dot{u}\|^2 = \tfrac{1}{2}\int |\dot{u}(x)|^2\,dx + \tfrac{1}{2}\int |u_{xx}(x)|^2\,dx,$$

(well-known to be equivalent to the $H^2 \times L_2$ norm). This is proved by exactly the same procedure as in the example of "panel flutter" given in Section 5.

Secondly, the nonlinear term

$$J(u,\dot{u}) = \begin{pmatrix} 0 \\ T(u_x)_x \end{pmatrix}$$

is a C^∞ map of X to X. Indeed, from the composition theorems given in Section 4, the following maps are all smooth (we recall that in one dimension, $H^2 \subset C^0)^\dagger$:

$$\begin{array}{ccccccc} H^2 & \longrightarrow & H^1 & \longrightarrow & H^1 & \longrightarrow & L_2, \\ u & \longmapsto & u_x & \longmapsto & T(u_x) & \longmapsto & (T(u_x))_x. \end{array}$$

It is a general (and rather simple) fact that a C^∞ perturbation of a linear generator generates a local semiflow and that this semiflow can be maximally extended in time. (See Segal [227]) Moreover, since J and its derivatives are bounded on bounded sets, the time of existence is uniformly

---

†Note that this sort of example "just" fails in two dimensions since $H^1 \not\subset C_0$. However, two dimensional models related to those in Holmes and Marsden [128] do work.

bounded away from zero on bounded sets. Also, from our work in Section 9, or directly, we know that energy is conserved along solutions. This shows that (S1) is valid.

Since $H^2 \subset W^{1,\infty}$, condition (S2)(a) is valid, as in 5(a) above. Also, since $V : H^2 \longrightarrow \mathbb{R}$ is smooth,

$$V(0) = 0, \quad DV(0) = 0,$$

and

$$D^2V(0)(v,w) = \int v_{xx}w_{xx}dx + \int W''(0)v_x w_x dx,$$

V has a non-degenerate critical point at zero (in the strong sense that $D^2V(0)$ yields an isomorphism of H^2 to H^2 which, in this case is $\lambda - \Delta$, where $\lambda = W''(0) \geqslant 0$). Therefore, by Taylor's theorem (or the Morse lemma, if one desires) one easily obtains (S2)(b).

This proves that in this model, the trivial solution lies in a potential well and that it is dynamically stable.†

One must admit, however, that the sense in which semilinear approximations approximate the dynamics is not understood. For instance, periodic orbits and a vibrational analysis may be rigorously justified using Hamiltonian methods for semilinear equations. However, as in McCamy and Mizel [179], periodic orbits are probably impossible in the usual semilinear elastic models because of shocks.

It is perhaps appropriate to end these notes with a simple sounding but non-trivial problem. Is there a precise way in which observed periodic vibrations (in rods, say) can be predicted from an "exact" nonlinear elastic model.

---

†If one assumes $W(u_x) \geqslant 0$ for all u_x, then the energy estimate shows that solutions are global in time for any initial data. We proved global existence near the trivial solution in remark 5 above.

REFERENCES

[1] Abraham, R. and J. Marsden, Foundations of Mechanics, Second Edition, Benjamin (1978).

[2] Agmon, S., Lectures on Elliptic Boundary Value Problems, D. Van Nostrand (1965).

[3] Agmon, S., A. Douglis and L. Nirenberg, Estimates near the boundary for solutions of elliptic partial differential equations satisfying general boundary conditions, Comm. Pure and Appl. Mathematics 12(1959), 35-92.

[4] Antman, S. S., The Theory of Rods, Handbuch der Physik VIa/2, (ed. C. Truesdell) Springer (1972).

[5] Antman, S. S., Ordinary differential equations of non-linear elasticity I: Foundations of the theories of non-linearly elastic rods and shells, Arch. Rat. Mech. Anal. 61(1976), 307-351.

[6] Antman, S. S., Ordinary differential equations of non-linear elasticity II: Existence and regularity theory for conservative boundary value problems, Arch. Rat. Mech. Anal. 61(1976), 353-393.

[7] Antman, S. S., Nonuniqueness of equilibrium states for bars in tension, J. Math. Anal. Appl. 44(1973), 333-349.

[8] Antman, S. S. and G. Rosenfeld, Global behaviour of buckled states of nonlinearly elastic rods, S.I.A.M. Review, July (1978).

[9] Arnold, V., Sur la geometrie differentielle des groupes de Lie de dimension infinie et ses applications à l'hydrodynamique des fluids parfaits, Ann. Inst. Fourier Grenoble 16(1966), 319-361.

[10] Balakrishnan, A. V., Applied Functional Analysis, Springer (1976).

[11] Ball, J. M., Saddle point analysis for an ordinary differential equation in a banach space, and an application to dynamic buckling of a beam, in Nonlinear Elasticity, (ed. J. Dickey), Academic Press, New York (1973), 93-160.

[12] Ball, J. M., Convexity conditions and existence theorems in nonlinear elasticity, Arch. Rat. Mech. Anal. 63(1977), 337-403.

[13] Ball, J. M., Strongly continuous semigroups, weak solutions and the variation of constants formula, Proc. AMS. 63(1977), 370-373.

[14] Ball, J. M., Remarks on blow-up and non-existence for non-linear wave equations, Quart. J. Math. 28(1977), 473-486; see also Proc. Symp. "Nonlinear Evolution Equations", Univ. of Wisconsin (to appear).

[15] Ball, J. M., Constitutive inequalities and existence theorems in nonlinear elastostatics, in Nonlinear Analysis and Mechanics, Heriot-Watt Symposium, (R. J. Knops, ed.), Vol. I, Pitman (1977).

[16] Ball, J. M., R. J. Knops and J. E. Marsden, Two examples in nonlinear elasticity, Proc. Conf. on Nonlinear Analysis, Bescancon (June 1977), Springer Lecture Notes in Mathematics (to appear).

[17] Beatty, M. F., On the foundation principles of general classical mechanics, Arch. Rat. Mech. Anal. (1967), 264-273.

[18] Beatty, M. F., Some static and dynamic implications of the general theory of elastic stability, Arch. Rat. Mech. Anal. 19(1965), 167-188.

[19] Beatty, M. F., A theory of elastic stability for incompressible, hyperelastic bodies, Int. J. Solids Structures, 3(1967), 23-37.

[20] Beatty, M.F., Stability of the undistorted states of an isotropic elastic body, Int. J. Nonlinear Mech. 3(1968), 337-349.

[21] Beevers, C. E., Uniqueness and stability in linear visco-elasticity, J. Appl. Math. Phys. (ZAMP), 26(1975), 177-186.

[22] Beevers, C. E., Continuous data-dependence results for a general theory of heat conduction in bounded and unbounded domains, Quart. of Appl. Math. 35(1977), 111-119.

[23] Beevers, C. E., Some stability results in the linear theory of anisotropic elastodynamics, J. of Elasticity, 6(1976), 419-428 (see also J. de Mecanique, 14(1975), 639-651).

[24] Beju, I., Theorems on existence, uniqueness and stability of the place boundary value problem, in statics, for hyperelastic materials, Arch. Rat. Mech. Anal. 42(1971), 1-23 (also Bull. Math. Soc. Roum. 16(1972), 132-149, 283-313).

[25] Berger, M. S., Applications of global analysis to specific nonlinear eigenvalue problems, Rocky Mtn. J. Math. 3(1973), 319-354.

[26] Bishop, R. and S. Goldberg, Tensor Analysis on Manifolds, Prentice Hall (1969).

[27] Blancheton, M., Mécanique analytique des milieux continus, Ann. Inst. H. Poincaré A 7(1967), 189-213.

[28] Bolza, O., Lectures on the Calculus of Variations, Chelsea, N.Y. (1973).

[29] Bramble, J. H. and L. E. Payne, Effects of measurement of elastic constants on the solutions of problems in classical elasticity, J. Res. NBS 67B(1963), 157-167.

[30] Brockway, G. S., On the uniqueness of singular solutions to boundary-initial value problems in linear elastodynamics, Arch. Rational Mech. Anal. 48(1972), 213-244.

[31] Browder, F., A priori estimates for solutions of elliptic boundary problems, Nederl. Akad. Wetensch. Indag. Math. 22(1960), 145-169, 23(1961), 404-410.

[32] Bourguignon, J. P., D. G. Ebin and J. E. Marsden, Sur le noyau des opérateurs pseudo différentiels à symbole surjective, et non-injective, C.R. Acad. Sci. Paris 282(1976), 867-870.

[33] Capriz, G. and P. Podio Guidugli, On Signorini's perturbation method in finite elasticity, Arch. Rat. Mech. Anal. 57(1974), 1-30.

[34] Carlson, D. E., Linear Thermoelasticity, Handbuch der Physik, BIa/2, (ed. C. Truesdell), Springer (1972).

[35] Carroll, M. M. and P. M. Naghdi, The influence of the reference geometry on the response of elastic shells, Arch. Rat. Mech. Anal., 48(1972), 302-318.

[36] Carter, B. and H. Quintana, Foundations of general relativistic high-pressure elasticity theory, Proc. Roy. Soc. London A331(1972), 57-83 (see also Comm. Math. Phys. 30(1973), 261-286).

[37] Caughey, T. K. and R. T. Shield, Instability and the energy criteria for continuous systems, Z. Angew. Math. Phys. 19(1968), 485-492.

[38] Chernoff, P. R., Essential self-adjointness of powers of generators of hyperbolic equations. J. Funct. Anal. 12(1973), 401-414.

[39] Chernoff, P. and J. Marsden, Some properties of infinite dimensional hamiltonian systems, Springer Lecture Notes 425(1974) (see also, On continuity and smoothness of group actions, Bull. Am. Math. Soc. 76(1970), 1044-1049).

[40] Chevallier, D. P., Structures de variétés bimodelées et dynamique analytique des milieux continus, Ann. Ins. H. Poinare A21(1974), 43-76.

[41] Chillingworth, D., The catastrophe of a buckling beam, Springer Lecture Notes in Math. 488(1975), 88-91.

[42] Chillingworth, D. and J. E. Marsden, The traction problem near equilibrium in finite elasticity (in preparation).

[43] Choquet-Bruhat, Y. (Y. Foures-Bruhat), Théorème d'existence pour certain systèmes d'équations aux dérivées partielles non linéaires, Acta Math. 88(1952), 141-225.

[44] Choquet-Bruhat, Y. and L. Lamoureau-Brousse, Sur les équations de l'elasticité relativiste, C.R. Acad. Sci. Paris 276(1973), 1317-1320.

[45] Choquet-Bruhat, Y. and J. E. Marsden, Solution of the local mass problem in general relativity, Comm. Math. Phys. 51(1976), 283-296.

[46] Chorin, A., T. Hughes, M. McCracken and J. E. Marsden, Product formulas and numerical algorithms. Comm. Pure and Appl. Math. 31(1978), 205-256.

[47] Chow, S., J. Hale and J. Mallet-Paret, Applications of generic bifurcation, I, II, Arch. Rat. Mech. Anal. 59(1975), 159-188, 62(1976), 209-235.

[48] Coleman, B. D., Thermodynamics of materials with memory, Arch. Rat. Mech. Anal. 17(1964), 1-16.

[49] Coleman, B. D., On the mechanics of materials with fading memory, Springer Lecture Notes in Math. 503(1976), 290-293.

[50] Coleman, B. D., On the energy criterion for stability, in Nonlinear Elasticity, (ed. R. W. Dickey), Academic Press (1974), 31-53.

[51] Coleman, B. D., and E. H. Dill, On the stability of certain motions of incompressible materials with memory, Arch. Rat. Mech. Anal. 30(1968), 197-224.

[52] Coleman, B. D., J. M. Greenberg and M. E. Gurtin, Waves in materials with memory V. On the amplitude of acceleration waves and mild discontinuities, Arch. Rat. Mech. Anal. 22(1966), 333-354.

[53] Coleman, B. D. and V. J. Mizel, Existence of entropy as a consequence of asymptotic stability, Arch. Rat. Mech. Anal. 24(1967) 243-270.

[54] Coleman, B. D. and V. J. Mizel, On thermodynamic conditions for the stability of evolving systems, Arch. Rat. Mech. Anal. 29(1968), 105-113.

[55] Coleman, B. D. and V. J. Mizel, Norms and semi-groups in the theory of fading memory, Arch. Rat. Mech. Anal. 23(1966), 87-123.

[56] Coleman, B. D. and W. Noll, The thermodynamics of elastic materials with heat conduction and viscosity, Arch. Rat. Mech. An. 13(1963), 167-178.

[57] Cook, J. M., Complex Hilbertian structures on stable linear dynamical systems, J. Math. and Mech. 16(1966), 339-349.

[58] Courant, R. and D. Hilbert, Methods of Mathematical Physics, Vol. II, Wiley: Interscience (1962).

[59] Crandall, M. G., The semigroup approach to first order quasi-linear equations in several space variables, Israel J. Math. 12(1972), 108-132.

[60] Croll, J. G. A. and A. C. Walker, Elements of Structural Stability, Macmillan (1972).

[61] Dafermos, C. M., Contraction semigroups and trend to equilibrium in continuum mechanics, Springer Lecture Notes in Math. 503(1976), 295-306.

[62] Dafermos, C. M., The mixed initial-boundary value problem for the equations of nonlinear one-dimensional viscoelasticity, J. Diff. Eqns. 6(1969), 71-86.

[63] Dafermos, C. M., On the existence and the asymptotic stability of solutions to the equations of linear thermo-elasticity, Arch. Rat. Mech. Anal. 29(1968), 241-271.

[64] Dafermos, C. M., Solution of the Riemann problem for a class of hyperbolic systems of conservation laws by the viscosity method, Arch. Rat. Mech. Anal. 52(1973), 1-9.

[65] Dafermos, C. M., Quasilinear hyperbolic systems, in Nonlinear Waves, (ed. J. Newell), Cornell Univ. Press (1974), 83-100.

[66] Dafermos, C. M. and M. Slemrod, Asymptotic behavior of nonlinear contraction semigroups, J. Functional Analysis, 13(1973), 97-106.

[67] Day, W. A., A condition equivalent to the existence of entropy in classical thermodynamics, Arch. Rat. Mech. Anal. 49(1972), 159-171.

[68] Dieudonné, J., Foundations of Modern Analysis, Academic Press (1960).

[69] Dionne, P., Sur les problèmes de Cauchy bien posés, J. Anal. Math. Jerusalem 10(1962/63), 1-90.

[70] Dorroh, J. R. and J. E. Marsden, Differentiability of Nonlinear Semigroups (Unpublished Notes)

[71] Dowell, E. H., Aeroelasticity of Plates and Shells, Nordhoff (1975).

[72] Dunford, N. and J. Schwartz, Linear Operators, Vol. II Interscience, Wiley (1963).

[73] Duvaut, T. G. and J. L. Lions, Les Inéquations en Mécanique et en Physique, Dunod (1972).

[74] Dyson, F. J., Missed opportunities, Bull. AMS 78(1972), 635-652.

[75] Ebin, D. G., The manifold of Riemannian metrics, Proc. Symp. Pure Math. 15(1970), 11-40.

[76] Ebin, D. G., The motion of slightly compressible fluids viewed as a motion with strong constraining force, Ann. of Math. 105(1977), 141-200.

[77] Ebin, D. G. and J. E. Marsden, Groups of diffeomorphisms and the motion of an incompressible fluid, Ann. of Math. 92(1970), 102-163.

[78] Ericksen, J. L., A thermokinetic view of elastic stability theory, Int. Journ. Solid. Structures, 2(1966), 573-580.

[79] Ericksen, J. L., Thermoelastic stability, Proc. Fifth. U.S. Congr. on Appl. Mech. (1966), 187-193.

[80] Ericksen, J. L., Non-existence theorems in linearized elastostatics, J. Diff. Eq'ns. 1(1965), 446-451.

[81] Eringen, A. C. (Ed), Continuum Physics, Vol. II, Academic Press (1975).

[82] Eringen, A. C. and E. S. Suhubi, Elastodynamics, Vol. I, II, Academic Press (1974/5).

[83] Fichera, G., Existence theorems in elasticity, Handbuch der Physik, Vol. IVa/2, (ed. C. Truesdell), Springer (1972).

[84] Fischer, A. and J. E. Marsden, The Einstein equations of evolution - a geometric approach, J. Math. Phys. 13(1972), 546-568.

[85] Fischer, A. and J. E. Marsden, The Einstein evolution equations as a first-order quasi-linear symmetric hyperbolic system, I, Comm. Math. Phys. 28(1972), 1-38.

[86] Fischer, A. and J. E. Marsden, Deformations of the scalar curvature, Duke Math. J. 42(1975), 519-547.

[87] Fischer, A. and J. E. Marsden, Linearization stability of nonlinear partial differential equations, Proc. Symp. Pure Math. 27(1975), Part 2, 219-263.

[88] Fischer, A. and J. E. Marsden, Hamiltonian field theories on space-time (in preparation)

[89] Fletcher, D. C., Conservation laws in linear elastodynamics, Arch. Rat. Mech. Anal. 60(1976), 329-353.

[90] Frankl, F., Über das Anfangswertproblem für lineare und nichtlineare hyperbolische partielle differentialgleichungen zweiter Ordnung, Mat. Sb. 2(44)(1937), 814-868.

[91] Friedman, A., Partial Differential Equations, Holt, Rinehart and Winston (1969).

[92] Friedrichs, K. O., On the boundary value problems of the theory of elasticity and Korn's inequality, Ann. of Math. 48(1947), 441-471.

[93] Friedrichs, K. O., Symmetric hyperbolic linear differential equations. Commun. Pure Appl. Math. 7(1954), 345-392.

[94] Friedrichs, K. O., Symmetric positive linear differential equations. Commun. Pure Appl. Math. 11(1958), 333-418.

[95] Gårding, L., Energy inequalities for hyperbolic systems in differential analysis, Bombay Colloq. (1964), see also Cauchy's Problem for Hyperbolic Equations, Lecture Notes, Chicago (1957).)

[96] Goldstein, J., Semigroups and second order differential equations, J. Funct. Anal. 4(1969), 50-70.

[97] Green, A. E. and J. E. Adkins, Large Elastic Deformations, Second Edition, Oxford Univ. Press (1970).

[98] Green, A. E. and P. M. Naghdi, On continuum thermodynamics, Arch. Rat. Mech. Anal. 48(1972), 352-378.

[99] Green, A. E., P.M. Naghdi and R. S. Rivlin, Directors and multipolar displacements in continuum mechanics, Int. J. Engng. Sci. 2(1965), 611-620.

[100] Green, A. E., P. M. Naghdi and J. A. Trapp, Thermodynamics of a continuum with internal constraints, Int. J. Eng. Sci. 8(1970), 891-908.

[101] Green, A. E. and R. S. Rivlin, On Cauchy's equations of motion, Z. Angew Math. Phys. 15(1964), 290-292.

[102] Green, A. E. and R. S. Rivlin, Multipolar continuum mechanics, Arch. Rat. Mech. Anal. 17(1964), 113-147.

[103] Green, A. E. and W. Zerna, Theoretical Elasticity, 2nd Edition, Oxford Univ. Press (1968).

[104] Greenberg, J. M., R. C. MacCamy and V. J. Mizel, On the existence, uniqueness and stability of solutions of the equation $\rho\chi_{tt} = E(\chi_x)\chi_{xx} + \lambda\chi_{xxt}$, J. Math. Mech. 17(1968), 707-728.

[105] Grioli, G., Mathematical Theory of Elastic Equilibrium, Ergebnisse der Ang. Mat. No.7, Springer (1962).

[106] Gurtin, M. E., The linear theory of elasticity, Handbuch der Physik, Vol. IVa/2, (ed. C. Truesdell), Springer (1972).

[107] Gurtin, M. E., Modern continuum thermodynamics, Mechanics Today, Vol.1, (ed. S. Nemat-Nasser)(1972), 168-213.

[108] Gurtin, M. E., On the thermodynamics of elastic materials, J. Math. Anal. and Appl. 18(1967), 38-44.

[109] Gurtin, M. E., An introduction to classical continuum mechanics Notes, Carnegie-Mellon University, Pittsburgh (1977).

[110] Gurtin, M. E., Thermodynamics and the energy criterion for stability, Arch. Rat. Mech. Anal. 52(1973), 93-103.

[111] Gurtin, M. E., Thermodynamics and stability, Arch. Rat. Mech. Anal. 59(1975), 63-96.

[112] Gurtin, M. E., and L. C. Martins, Cauchy's theorem in classical physics, Arch. Rat. Mech. Anal. 60(1976), 305-324, 325-328.

[113] Gurtin, M. E. and A. Ian Murdoch, A continuum theory of elastic material surfaces, Arch. Rat. Mech. Anal. 57(1975), 291-323; 59(1975), 389-390.

[114] Gurtin, M. E., and E. Sternberg, A note on uniqueness in classical elastodynamics, Quart. Appl. Math. 19(1961), 169-171.

[115] Gurtin, M. E. and R. A. Toupin, A uniqueness theorem for the displacement boundary value problem of linear elastodynamics, Quart. Appl. Math. 23(1965), 79-81.

[116] Gurtin, M. E. and W. O. Williams, On the Clausius-Duhem inequality, Z. Angew. Math. Phys. 17(1966), 626-633.

[117] Gurtin, M. E. and W. O. Williams, On the first law of thermodynamics, Arch. Rat. Mech. Anal. 42(1971), 77-92.

[118] Hale, J., Generic Bifurcation Theory, Nonlinear Analysis and Mechanics, Heriot-Watt Symposium, (R. J. Knops, ed.), Vol. I, Pitman (1977).

[119] Hawking, S. W. and G. F. R. Ellis, The Large Scale Structure of Spacetime, Cambridge University Press (1973).

[120] Hayes, M. and R. J. Knops, On the displacement boundary-value problem of linear elastodynamics, Quart. of Appl. Math. 26(1968), 291-293.

[121] Hermann, R., Vector Bundles in Mathematical Physics I, II. W. A. Benjamin (1971).

[122] Hermann, R., Interdisciplinary Mathematics Volumes 4, 5 and 11, Rutgers University (1975).

[123] Herrmann, G., (ed.) Dynamic Stability of Structures, Pergamon (1966).

[124] Hill, R., On uniqueness and stability in the theory of finite elastic strain, J. Mech. Phys. Solids 5(1957), 229-241.

[125] Hille, E., and R. Phillips, Functional Analysis and Semi-groups, AMS Colloquium Publications (1967).

[126] Hills, R. N. and R. J. Knops, Continuous dependence for the compressible linear viscous fluid, Arch. Rat. Mech. Anal. 51(1973), 54-59.

[127] Hills, R. N. and R. J. Knops, Concavity and the evolutionary properties of a class of general materials, Proc. Roy. Soc. Edin. 72(1973/4), 239-243.

[128] Holmes, P. and J. Marsden, Bifurcations of Dynamical Systems and Nonlinear Oscillations in Engineering Systems, Proc. Conf. on Nonlinear PDE, Bloomington, Indiana, Nov.(1976) (ed. J. Chadam), Springer Lecture Notes, 648(1978), (see also Automatica, June(1978)).

[129] Hughes, T. J. R., T. Kato and J. E. Marsden, Well-posed quasi-linear second-order hyperbolic systems with applications to nonlinear elastodynamics and general relativity, Arch. Rat. Mech. Anal. 63(1977), 273-294.

[130] Hughes, T. J. R. and J. E. Marsden, Some applications of geometry in continuum mechanics, Rep. on Math. Phys. 12(1977), 35-44.

[131] Hughes, T. J. R. and J. E. Marsden, Classical elastodynamics as a symmetric hyperbolic system, J. of Elasticity, 8(1978), 97-110.

[132] Hughes, T. J. R. and J. E. Marsden, Mathematical Foundations of continuum mechanics, Notes for Math. 275-290, Berkeley (1976).

[133] Hughes, T. J. R. and J. E. Marsden, A Short Course in Fluid Mechanics, Publish or Perish (1976)

[134] Infante, E. F. and M. Slemrod, An invariance principle for dynamical systems on a Banach space: application to the general problem of thermoelastic stability. Symposium on Instability of Continuous Systems, (ed. H. Leipholz), Springer (1971), 215-221.

[135] John, F., Plane strain problems for a perfectly elastic material of harmonic type, Comm. Pure and Appl. Math. 13(1960, 239-296.

[136] Kato, T., Perturbation Theory for Linear Operators, Springer (1966).

[137] Kato, T., Linear evolution equations of hyperbolic type I, J. Fac. Sci. Univ. Tokyo 17(1970), 241-258.

[138] Kato, T., Linear evolution equations of hyperbolic type, II, J. Math. Soc. Japan 25(1973), 648-666.

[139] Kato, T., The Cauchy problem for quasi-linear symmetric hyperbolic systems, Arch. Rat. Mech. Anal. 58(1975), 181-205.

[140] Kato, T., Quasi-linear Equations of Evolution, with Applications to Partial Differential Equations, Springer Lecture Notes 448(1975), 25-70.

[141] Kato, T., On an inequality of Hardy, Littlewood and Polya, Advances in Mathematics 7(1971), 217-218.

[142] Kato, T., Linear and quasilinear equations of evolution of hyperbolic type, CIME Bressanone (1976).

[143] Knops, R. J. (ed.), Symposium on Non-well Posed Problems and Logarithmic Convexity, Springer Lecture Notes in Math. 316(1973).

[144] Knops, R. J. (ed.), Nonlinear Analysis and Mechanics; Heriot-Watt Symposium, Vol. I, Pitman (1977)

[145] Knops, R. J., Comments on nonlinear elasticity and stability, in Ordinary and Partial Differential Equations, Dundee (1976), (ed. by W. N. Everitt and B. D. Sleeman), Springer Lecture Notes in Mathematics 564(1976), 271-290.

[146] Knops, R. J., On potential wells in nonlinear elasticity. Some Aspects of Mechanics of Continua. Part 1, Sen Memorial Volume 72-86. Calcutta (1977).

[147] Knops, R. J., H. A. Levine and L. E. Payne, Non-existence, instability and growth theorems for solutions of a class of abstract nonlinear equations with applications to nonlinear elastodynamics, Arch. Rat. Mech. 55(1974), 52-72.

[148] Knops, R. J. and L. E. Payne, Growth estimates for solutions of evolutionary equations in Hilbert space with applications to elastodynamics, Arch. Rat. Mech. Anal. 41(1971), 363-398.

[149] Knops, R. J. and L. E. Payne, Hölder stability and logarithmic convexity, in Instability of Continuous Systems, IUTAM Symp. (1969), (ed. H. Leipholtz), Springer (1971).

[150] Knops, R. J. and L. E. Payne, Continuous data dependence for the equations of classical elastodynamics, Proc. Camb. Phil. Soc. 66(1969), 481-491.

[151] Knops, R. J. and L. E. Payne, Uniqueness Theorems in Linear Elasticity, Springer (1971).

[152] Knops, R. J. and L. E. Payne, Uniqueness in classical elastodynamics, Arch. Rat. Mech. Anal. 27(1968), 349-355.

[153] Knops, R. J. and L. E. Payne, On uniqueness and continuous dependence in dynamical problems of linear thermoelasticity, Int. J. Solids Structures 6(1970), 1173-1184.

[154] Knops, R. J. and L. E. Payne, On potential wells and stability in nonlinear elasticity, Math. Proc. Camb. Phil. Soc. (to appear).

[155] Knops, R. J. and B. Straughan, Non-existence of global solutions to non-linear Cauchy problems arising in Mechanics. Trends in Applications of Pure Mathematics to Mechanics, (ed. G. Fichera), Pitman (1976).

[156] Knops, R. J. and B. Straughan, Convergence of solutions of the equations of dynamic linear dipolar elasticity to the solutions of classical elastodynamics, Archives of Mech. (Warszawa) 28(1976), 431-441.

[157] Knops, R. J. and E. W. Wilkes, Theory of Elastic Stability, Handbuch der Physik (ed. C. Truesdell), Vol. VIa/3, Springer (1973).

[158] Knowles, J. K. and E. Sternberg, On a class of conservation laws in linearized and finite elasticity, Arch. Rat. Mech. Anal. 44(1972), 187-211.

[159] Knowles, J. K. and E. Sternberg, On the ellipticity of the equations of nonlinear elastostatics for a special material, J. of Elasticity 5(1975), 334-341.

[160] Knowles, J. K. and E. Sternberg, On the failure of ellipticity of the equations for finite elastostatic plane strain, Arch. Rat. Mech. Anal. 63(1977), 321-336.

[161] Kobelkov, G. M., On existence theorems for some problems in the theory of elasticity, Mat. Zametki 17(1975), 599-609.

[162] Koiter, W. T., Thermodynamics of elastic stability, Third Can. Congr. on Appl. Mech. (1971), 29-37.

[163] Koiter, W. T., On the instability of equilibrium in the absence of a minimum of the potential energy, Proc. Kon. Ned. Akad. Wet. B68(1965), 107-113.

[164] Koiter, W. T., The energy criterion of stability for continuous elastic bodies, Proc. Kon. Ned. Ak. Wet. B68(1965), 178-202.

[165] Koiter, W. T., A basic open problem in the theory of elastic stability, Springer Lecture Notes in Math. 503(1976), 366-373.

[166] Krzyzanski, M. and J. Schauder, Quasilineare differentialgleichungen zweiter ordnung vom hyperbolischen Typus, Gemischte Randwertaufgaben, Studia. Math. 5(1934), 162-189.

[167] Lang, S., Differential Manifolds, Addison Wesley (1972).

[168] Lax, P. D., Cauchy's problem for hyperbolic equations and the differentiability of solutions of elliptic equations, Comm. Pure and Appl. Math. 8(1955), 615-633.

[169] Lax, P. D., Hyperbolic systems of conservation laws and the Mathematical Theory of Shock Waves, SIAM (1973).

[170] Lax, P. D., Development of singularities of solutions of nonlinear hyperbolic partial differential equations, J. Math. Phys. 5(1964), 611-613.

[171] Lax, P. D. and R. S. Phillips, Local boundary conditions for dissipative symmetric linear differential operators. Comm. Pure Appl. Math. 13(1960), 427-455.

[172] Leray, J., Hyperbolic Differential Equations, Institute for Advanced Study, Notes (1953).

[173] Lianis, G. and R. S. Rivlin, Relativistic equations of balance in continuum mechanics, Arch. Rat. Mech. Anal. 48(1972), 64-82.

[174] Lichnerowicz, A., Relativistic Hydrodynamics and Magnetohydrodynamics, Benjamin (1967).

[175] Littman, W., $L^p - L^q$ estimates for singular integral operators arising from hyperbolic equations. Proc. Symp. Pure Math. AMS 23(1973), 477-482.

[176] Lumer, G. and R. S. Phillips, Dissipative operators in a Banach space, Pac. J. Math. 11(1961), 679-698.

[177] MacCamy, R. C., Existence, Uniqueness and stability of $u_{tt} = \frac{\partial}{\partial x}(\sigma(u_x) + \lambda(u_x)u_{xt})$, Indiana Univ. Math. J. 20(1970), 231-238.

[178] MacCamy, R. C., A model for one-dimensional, nonlinear elasticity, Appl. Math. 35(1977), 21-33.

[179] MacCamy, R. C. and V. J. Mizel, Existence and nonexistence in the large of solutions of quasilinear wave equations, Arch. Rat. Mech. Anal. (1967), 299-320

[180] Marsden, J. E., Darboux's theorem fails for weak sympletic forms, Proc. Am. Math. Soc. 32(1972), 590-592.

[181] Marsden, J. E., Applications of Global Analysis in Mathematical Physics, Publish or Perish (1974).

[182] Marsden, J. E., On global solutions for nonlinear Hamiltonian evolution equations, Comm. Math. Phys. 30(1973), 79-81.

[183] Marsden, J. E., Qualitative methods in bifurcation theory, Bull. Am. Math. Soc. (to appear).

[184] Marsden, J. E., D. Ebin and A. Fischer, Diffeomorphism Groups, Hydrodynamics, and General Relativity, Proc. 13th Biennial Seminar of the Canadian Math. Congress, (ed. J. Vanstone), Montreal (1972), 135-279.

[185] Marsden, J. E. and M. McCracken, The Hopf Bifurcation and its Applications, Springer Notes in Applied Math. Sciences 19(1976).

[186] Marsden, J. E. and A. Weinstein, Reduction of symplectic manifolds with symmetry, Rep. on Math. Phys. 5(1974), 121-130.

[187] Massey III, F. J., Abstract evolution equations and the mixed problem for symmetric hyperbolic systems. Trans. Amer. Math. Soc. 168(1972), 165-188.

[188] Maugin, G., Sur la formulation des lois de comportement en mecanique relativiste des milieux continus, These, Univ. Paris VI (1975) (see also Ann. Inst. H. Poincaré 15(1971). 275-302, J. Gen. Rel. Grav. 4(1973), 241-272, and Comm. Math. Phys. 53(1977), 233-256.)

[189] Miller, R. K., An integro-differential equation for rigid heat conductors with memory, J. Math. Anal. Appl. (to appear).

[190] Mizohata, S., The Theory of Partial Differential Equations, Cambridge Univ. Press (1973).

[191] Morrey, C. B., Multiple Integrals in the Calculus of Variations, Springer (1966)

[192] Muller, I., A possible experiment on the principle of objectivity. Modern Developments in Thermodynamics, (ed. Benjamin Gal-Or) Wiley (1974).

[193] Murray, A. C., Uniqueness and continuous dependence for the equations of elastodynamics without strain energy function. Arch. Rat. Mech. Anal. 47(1972), 195-204.

[194] Naghdi, P. M., The Theory of Shells, Handbuch der Physik VIa/2 (ed. C. Truesdell) Springer (1972).

[195] Naghdi, P. M., and J. Trapp, On the general theory of stability for elastic bodies, Arch. Rat. Mech. Anal. 51(1973), 165-191.

[196] Nagy, Sz., Acta Sci. Math. Zzeged, 11(1947), 152-157 (see Riesz-Nagy [219] p.396).

[197] Navarro, C., Asymptotic stability in linear thermovisco-elasticity, J. Math. Anal. Appl. (to appear).

[198] Nelson, E., Analytic vectors, Ann. of Math. 70(1959), 572-615.

[199] Nirenberg, L., On elliptic partial differential equations, Ann. de Scuola. Norm. Sup. Pisa 13(1959), 115-162.

[200] Nirenberg, L., Topics in Nonlinear Analysis, Courant Institute (1974).

[201] Noll, W., A mathematical theory of the mechanical behaviour of continuous media, Arch. Rat. Mech. Anal. 2(1958), 197-226.

[202] Noll, W., A new mathematical theory of simple materials, Arch. Rat. Mech. An. 48(1972), 1-50.

[203] Noll, W., La mécanique classique basée sur un axiome d'objectivité, in La Méthode axiomatique dans les mécaniques classiques et nouvelles, Paris (1963), 47-56.

[204] Noll, W., Lectures on the foundations of continuum mechanics and thermodynamics, Arch. Rat. Mech. Anal. 52(1973), 62-92.

[205] Oden, J. T. and J. N. Reddy, Variational Methods in Theoretical Mechanics, Springer (1976).

[206] Palais, R. S., Seminar on the Atiyah-Singer Index Theorem. Ann. of Math. Studies, No.57, Princeton Univ. Press (1965).

[207] Palais, R., Foundations of Global Nonlinear Analysis, Benjamin (1968).

[208] Parks, P. C., A stability criterion for a panel flutter problem via the second method of Liapunov, "Differential Equations and Dynamical Systems", (ed. J. K. Hale and J. P. La Salle), Academic Press (1966).

[209] Payne, L. E. and D. H. Sattinger, Saddle points and instability of nonlinear hyperbolic equations, Israel J. Math. 22(1975), 273-303.

[210] Payne, L. E. and H. F. Weinberger, On Korn's inequality, Arch. Rat. Mech. Anal. 8(1961), 89-98.

[211] Pazy, A., Semi-groups of linear operators and applications to partial differential equations, Univ. of Maryland, Lect. Notes No.10, (1974).

[212] Pearson, C., General theory of elastic stability, Q. Appl. Math. 14(1956), 133-144.

[213] Petrovskii, I., Über das Cauchysche Problem für lineare und nicht-lineare hyperbolische partielle differentialgleichunger, Mat. Sb. 2(44)(1937), 814-868.

[214] Quinn, B., Solutions with shocks, an example of an L_1-contractive semigroup, Comm. Pure and Appl. Math. 24(1971), 125-132.

[215] Rabinowitz, P., Some global results for nonlinear eigenvalue problems, J. Funct. An. 7(1971), 487-513.

[216] Rauch, J. B. and F. J. Massey III, Differentiability of solutions to hyperbolic initial-boundary value problems, Trans. Amer. Math. Soc. 89(974), 303-3 8.

[217] Reed, M. and B. Simon, Methods of Modern Mathematical Physics II Fourier Analysis and Self Adjointness, Academic Press (1975).

[218] Reiss, E. L. and B. J. Matkowsky, Nonlinear dynamic buckling of a compressed elastic column, Appl. Math. 29(1971), 245-260.

[219] Riesz, F. and B. Sz. Nagy, Functional Analysis, Ungar (1955).

[220] Rivlin, R. S., Some topics in finite elasticity. Structural Mechanics, Pergamon (1960), 169-198.

[221] Rivlin, R. S., The fundamental equations of nonlinear continuum mechanics. Dynamics of Fluids and Plasmas, Academic Press, (1966), 83-126.

[222] Rostamian, R., Transition from unconstrained to constrained materials in continuum mechanics, Thesis, Brown Univ. (1978).

[223] Rubin, H. and P. Ungar, Motion under a strong constraining force, Comm. Pure and Appl. Math. 10(1957), 65-87.

[224] Schauder, J., Das Anfangswertproblem einer quasi-linearen hyperbolischen differentialgleichung zweiter ordnung in beliebiger Anzahl von unabhängigen Veränderlichen, Fund. Math. 24(1935), 213-246.

[225] Schwartz, Nonlinear Functional Analysis, Gordon and Breach, (1969).

[226] Segal, I., Symplectic structures and the quantization problem for wave equations, Sympt. Math. XIV. Academic Press (1974), 99-117.

[227] Segal, I., Nonlinear semigroups, Ann. of Math. 78(1963), 339-364.

[228] Signorini, A., Sulle deformazioni termoelastiche finite, Proc. 3rd Int. Cong. Appl. Mech. 2(1930), 80-89.

[229] Signorini, A., Transformazioni termoelastiche finite, Mem 1, Ann. di Mat. 22(1943), 33-143.

[230] Slemrod, M., Asymptotic behaviour of C_0 semigroups as determined by the spectrum of the generator, Indiana Univ. Math. J. 25(1976), 783-792.

[231] Sobolev, S. S., Applications of Functional Analysis in Mathematical Physics, Translations of Math. Monographs 7, Am. Math. Soc. (1963).

[232] Sobolev, S. S., On the theory of hyperbolic partial differential equations, Mat. Sb. 5(47)(1939), 71-99.

[233] Souriau, J. M., In Geometric Symplectique et Physique Mathematique, (ed. J. M. Souriau), Publ. No.235 du CNRS (1976).

[234] Spencer, A. J. M., Finite deformations of an almost incompressible elastic solid, Proc. Int'l. Symp. on Second Order Effects. Harfa (1962), Pergamon (1964).

[235] Stoppelli, F., Sulla sviluppibilità in serie di potenze di un parametro delle soluzioni delle equazioni dell' elastostatica isoterma, Recerche Mat. 4(1955), 58-73.

[236] Stoppelli, F., Sull' esistenza di soluzioni delle equazioni dell' elastostatica isoterma nel case di sollectizioni dotate di assi di equilibrio I, II, III, Richerche Mat. 6(1957), 241-287, 7(1958), 71-101, 7(1958), 138-152.

[237] Tartar, L., Variational Methods and Monotonicity, Univ. of Wisconsin, MRC Report, No.1571 (1974).

[238] Terng, C-L., Natural Vector Bundles and Natural Differential Operators, Thesis, Brandeis (1976) (see also R. S. Palais and C-L Terng, Natural Bundles have Finite Order, Topology (1977)).

[239] Ting, T. W., Problem of compatability and orthogonal decomposition of second order symmetric tensors in a compact Riemannian manifold with boundary, Arch. Rat. Mech. Anal. 64(1977), 221-243.

[240] Thompson, J. M. T. and G. W. Hunt, A General Theory of Elastic Stability, Wiley, Interscience (1973) and Towards a Unified Bifurcation Theory, Zeit. Ang. Math. Phys. 26(5)(1975), 581-604.

[241] Tromba, A. J., Almost Riemannian structures on Banach manifolds, the Morse lemma and the Darboux theorem, Can. J. Math. 28(1976), 640-652.

[242] Tromba, A. J., A general approach to Morse Theory, J. Diff. Geom. Jan. (1978) and Mem. Am. Math. Soc. No.194 (1977).

[243] Trotter, H. F., Approximation of semi-groups of operators, Pac. J. Math. 8(1958), 187-919.

[244] Trotter, H. F., On the product of semigroups of operators, Proc. AMS 10(1959), 545-551.

[245] Truesdell, C., The Elements of Continuum Mechanics, Springer (1966).

[246] Truesdell, C., Rational Thermodynamics, Springer (1965).

[247] Truesdell, C., An Introduction to Rational Continuum Mechanics, Vol. I, Academic Press (1977).

[248] Truesdell, C. and W. Noll, The Nonlinear Field Theories of Mechanics, III/3, Handbuch der Physik, (ed. S. Flügge), Springer (1965).

[249] Truesdell, C. and R. A. Toupin, The Classical Field Theories, Handbuch der Physik, III/1, (ed. S. Flügge), Springer (1960).

[250] Vainberg, M. M., Variational Methods in the Theory of Nonlinear Operators, Holden-Day (1964).

[251] Van Buren, M., On the existence and uniqueness of solutions to boundary value problems in finite elasticity. Thesis. Carnegie-Mellon University (1968). Westinghouse Research Laboratories Report 68-107-MEKMARI (1968).

[252] Wang, C.- C. and C. Truesdell, Introduction to Rational Elasticity, Noordhoff (1973).

[253] Weiss, B., Abstract vibrating systems, J. Math. and Mech. 17(1967), 241-255.

[254] Wells, R., Analysis on Real and Complex Manifolds, Prentice Hall (1974).

[255] Wheeler, L., A uniqueness theorem for the displacement problem in finite elastodynamics, Arch. Rat. Mech. Anal. 63(1977), 183-190.

[256] Wheeler, L. and R. R. Nachlinger, Uniqueness theorems for finite elastodynamics, J. Elasticity 4(1974), 27-36.

[257] Wilcox, C. H., Wave operators and asymptotic solutions of wave propagation problems of classical physics, Arch. Rat. Mech. Anal. 22(1966), 37-78.

[258] Wilkes, N. S., Thermodynamic restrictions on viscoelastic materials, Quart. of Mech. and Appl. Math. 30(1977), 209-222.

[259] Wilkes, N. S., Continuous dependence and instability in linear thermoelasticity (to appear).

[260] Wilkes, N. S., Continuous dependence and instability in visco-elasticity, J. de Mecanique (to appear).

[261] Wilkes, N. S., On the non-existence of semi-groups for some equations of continuum mechanics (to appear).

[262] Yosida, K., Functional Analysis, Springer (1971).

[263] Zeeman, E. C., Euler Buckling, Springer Lecture Notes in Math., 525(1976), 373-395.

[264] Ziegler, H., On the concept of elastic stability, Advances in Applied Mechanics, 4(1956), 351-403.

[265] Ziegler, H., Principles of Structural Stability, Blaisdell (1968)

Professor J. E. Marsden, Department of Mathematics, University of California, Berkeley, California 94720, U.S.A.

Professor T. J. R. Hughes, Division of Engineering and Applied Science, California Institute of Technology, Pasadena, California 91125, U.S.A.